THE COMPLETE GUIDE TO

MASONRY & STONEWORK

Updated with New Products & Techniques

- Poured Concrete
- Brick & Block
- Natural Stone
- Stucco

Creative Publishing international

MINNEAPOLIS, MINNESOTA
www.creativepub.com

Creative Publishing international

Creative Publishing international, Inc.
400 First Avenue North, Suite 300
Minneapolis, Minnesota 55401
1-800-328-0590
www.creativepub.com

Printed in China

10 9 8 7 6 5 4 3 2 1

The complete guide to masonry & stonework. -- 3rd ed.
p. cm.
Includes index.
At head of title: Black & Decker.
"Poured concrete, brick & block, natural stone, stucco."
Summary: "Includes traditional techniques for laying concrete, as well as new materials and techniques, such as tumbled concrete pavers, acid-etching for colored concrete slabs,and important green paving options, such as rain garden arroyos and permeable pavers"--Provided by
publisher.
ISBN-13: 978-1-58923-520-5 (soft cover)
ISBN-10: 1-58923-520-7 (soft cover)
1. Stonemasonry. I. Black & Decker Corporation (Towson, Md.) II. Title: Complete guide to masonry and stonework.

TH5401.C655 2010
693'.1--dc22

2009048793

The Complete Guide to Masonry & Stonework
Created by: The Editors of Creative Publishing international, Inc. in cooperation with Black & Decker. Black & Decker® is a trademark of The Black & Decker Corporation and is used under license.

President/CEO: Ken Fund

Home Improvement Group

Publisher: Bryan Trandem
Managing Editor: Tracy Stanley
Senior Editor: Mark Johanson
Editor: Jennifer Gehlhar

Creative Director: Michele Lanci-Altomare
Art Direction/Design: Jon Simpson, Brad Springer, James Kegley

Lead Photographer: Joel Schnell
Set Builder: James Parmeter
Production Managers: Linda Halls, Laura Hokkanen

Page Layout Artist: Parlato Design Studio
Shop Help: Charles Boldt
Edition Editor: Kristen Hampshire
Copy Editor: Mark Kakkuri
Proofreader: Leah Noel

NOTICE TO READERS

For safety, use caution, care, and good judgment when following the procedures described in this book. The Publisher and Black & Decker cannot assume responsibility for any damage to property or injury to persons as a result of misuse of the information provided.

The techniques shown in this book are general techniques for various applications. In some instances, additional techniques not shown in this book may be required. Always follow manufacturers' instructions included with products, since deviating from the directions may void warranties. The projects in this book vary widely as to skill levels required: some may not be appropriate for all do-it-yourselfers, and some may require professional help.

Consult your local Building Department for information on building permits, codes and other laws as they apply to your project.

Contents

The Complete Guide to Masonry & Stonework

17 18 36 44 56 62

72

80

94

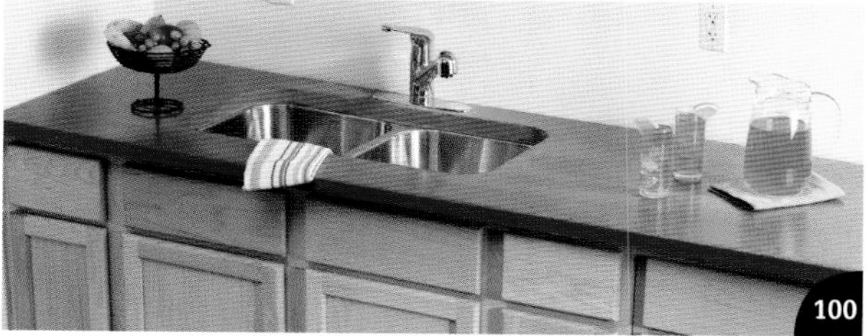

100

110

138

Contents (Cont.)

150

156

164

170

194

182

214

219

234

Contents (Cont.)

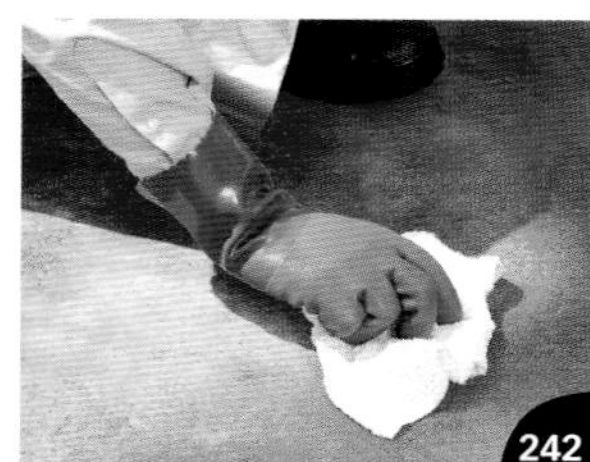

Introduction

Masonry is a popular home building material for many reasons, including its beauty, versatility, and resistance to fire, earthquakes, and sound transmission. And let's not forget its remarkable durability. While few of us imagine that our homes will exist for centuries, when we choose masonry we're choosing a material that has precisely that capability. Some of the world's most venerable masonry structures—the Taj Mahal, the Egyptian Pyramids, the Colosseum in Rome, the Great Sphinx of Giza—have awed countless generations with their ability to withstand time.

Poured Concrete introduces you to the tools, materials, and basic techniques necessary for accomplishing functional and attractive concrete projects. You'll learn foundation skills that allow you to build projects ranging from foundations to retaining walls and for inside, a kitchen island countertop. We walk you through each project, step-by-step, providing illustrative photos to guide the way.

Brick & Block includes projects such as building a brick barbecue and laying a mortared brick patio. You'll learn the tools, materials, and skills necessary to complete basic brick projects and sophisticated designs. A comprehensive section of projects gives you an opportunity to explore the possibilities of brick and block as a functional, long-lasting, and aesthetically pleasing construction medium.

Natural Stone introduces fresh design interpretations for an age-old material. Versatile and texturally interesting, natural stone is an increasingly popular landscaping material, and a robust selection of paver materials that resemble old-world stone are accessible and easy to use in building projects. We'll show you the various types of stone, how to choose appropriate material for your project, and techniques you'll want to master before taking on the projects in this book. Then, we give you a portfolio of natural stone ideas and step-by-step instructions to accomplish them at home.

Decorative Masonry Finishing shows you how to add interest to concrete surfaces, install veneer siding, and accomplish a stucco finish. These skills will help you customize other projects in this book.

Repair & Maintenance teaches you how to make quick fixes and take on more substantial repair projects to keep your masonry and stonework projects in top condition.

So, dig in! Peruse each section and earmark projects that interest you. You may want to make a priority list—one project inevitably leads to another. Consider yourself a masonry apprentice learning a time-honored craft and let us expose you to the hands-on techniques and creative possibilities.

Gallery of Masonry & Stonework

In their raw state, concrete and rocks are about the most humble building materials you can find. But add some inspiration and a little hard work, and you'll be amazed at the feats that can be accomplished with these simple products. The projects featured on the following pages are just a sampling of the beauty of masonry.

Stone slabs placed in a slope create natural garden steps in casual or formal settings.

A natural stone retaining wall and mortared flagstone driveway add structural interest to this home and provide a formal entryway.

Natural clay tile and a clay brick garden wall are separated by a border of light porcelain tiles. The contrasting porcelain color creates a visual bridge between the slightly differing brick tones of the patio and wall.

This stacked stone wall provides the framework for an outdoor kitchen. It combines natural and cast stones.

Stucco is a very popular and highly durable siding product that is created using modified masonry materials, tools, and techniques.

Blocks, pavers, and natural stone all find a home in this landscape. Included are two retaining walls (one cut stone, one interlocking block with cast capstones), a concrete paver walkway, and well-chosen landscape boulders.

Terracing is a landscaping technique in which a series of retaining walls are built to break up a slope into small, flat areas.

Cast cobble stones emulate the look of natural stone paving, with tumbled surfaces and random coloration. They are usually laid in regular patterns.

Cast concrete can be formed into a host of useful and decorative items for the garden, such as this planter cast in a 5-gallon bucket.

Poured concrete sidewalks don't need to be straight and uniform. Curves go a long way toward visually softening this rock-hard material.

Veneer stone has a refined, timeless appearance that is a very convincing imitation of natural stone. Because the manmade product weighs less than natural stone and is engineered for ease of installation, it is a very practical solution for siding your home.

Brick veneer transforms these cast concrete column bases from unremarkable to elegant. The resulting posts have the strength and durability of concrete and the beauty of brick.

The massive presence of concrete is used to great design advantage in the cast concrete elements found in the patio and fireplace above. The simple lines and monochromatic tones give the patio a contemporary feel.

Concrete pavers in various sizes and shapes add visual appeal when used in stairs and as planters.

Cast stepping stones allow landscapers to apply creative touches to the otherwise plain concrete gray tones. In this case, the gaps between stepping stones are filled with crushed gravel.

The exposed aggregate squares in this concrete patio take on the look of high-end landscape design because they are separated into a grid bordered by brick pavers.

Curves and nonlinear designs add great visual interest and create natural stopping points near landscape features.

A stairway landing is used to create a resting place and to allow you to adjust the stairway design so all of the treads are the same depth.

Cut flagstone patios and walkways are more formal in appearance and easier to walk on than surfaces made with uncut flags.

BLACK & DECKER

Poured Concrete

Poured concrete is versatile, sturdy, and highly functional for walkways, foundations, and even decorative features like garden steps. Concrete is back-to-the-basics—a standard material because of its weather-ready qualities and ability to stand the test of time. Plus, with today's creative applications for poured concrete you'll enjoy customizing a slab, experimenting with casting, and learning how to pour structures that will enhance your living environment.

In this chapter:

Poured concrete might not be the first material that comes to mind when dreaming of a new garden wall, but it's certainly worth consideration. The versatility of concrete can inspire all sorts of custom creations, such as this retaining wall with a traditional frame-and-panel effect.

Concrete Basics

Durable, versatile, and economical, poured concrete can be shaped and finished into a wide variety of surfaces and structures throughout your home and yard. Decorative surfaces and unique appearances can be produced with exposed aggregate, tints and stains, and special stamping tools.

Pouring footings or a foundation is the first step to many hardscape projects, such as fences, walls, sheds, and decks or gazebos, and is an excellent introduction to working with concrete. Pouring a small sidewalk or patio is a good way to learn about finishing concrete. It's best to save larger projects, such as driveways and large patios, until you are comfortable working with and finishing concrete and have plenty of assistants on hand to help you.

Planning and preparation are the keys to successful concrete projects. Poured concrete yields the most durable and attractive final finish when it is poured at an air temperature between 50 and 80 degrees F and when the finishing steps are completed carefully in the order described in the following pages.

Good preparation means fewer delays at critical moments and leaves you free to focus on placing and smoothing the concrete—not on staking loose forms or locating misplaced tools. Before beginning to mix or pour the concrete, make sure the forms are sturdy enough to stand up to the weight and pressure that will be exerted on them and that they are staked and braced well. Forms that are taller than four or five inches should be tied with wire. The joints on the forms should be tight enough that the bleed water doesn't run through them.

One of the most difficult aspects of finishing concrete is recognizing when it's ready. Many people try to rush the process, with disappointing results. Wait until the bleed water disappears and the concrete has hardened somewhat before floating the surface. A good rule of thumb is when the footprints you leave are light enough that you can no longer identify the type of shoes you are wearing, the concrete is ready to be worked.

Components of Concrete ▸

The basic ingredients of concrete are the same, whether the concrete is mixed from scratch, purchased premixed, or delivered by a ready-mix company. Portland cement is the bonding agent. It contains crushed lime, cement, and other bonding minerals. Sand and a combination of aggregates add volume and strength to the mix. Water activates the cement, and then evaporates, allowing the concrete to dry into a solid mass. By varying the ratios of the ingredients, professionals can create concrete with special properties that are suited for specific situations.

Tools & Materials

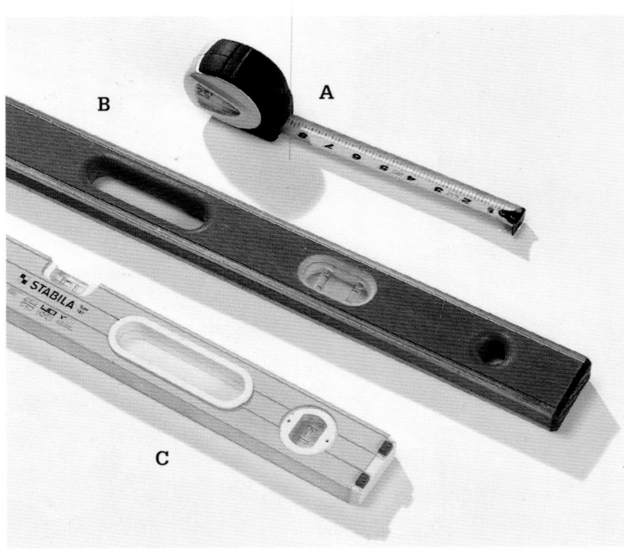

Layout and measuring tools for preparing jobsites and installing and leveling concrete forms include a tape measure (A), a 4-ft. level (B), and a 2-ft. level (C).

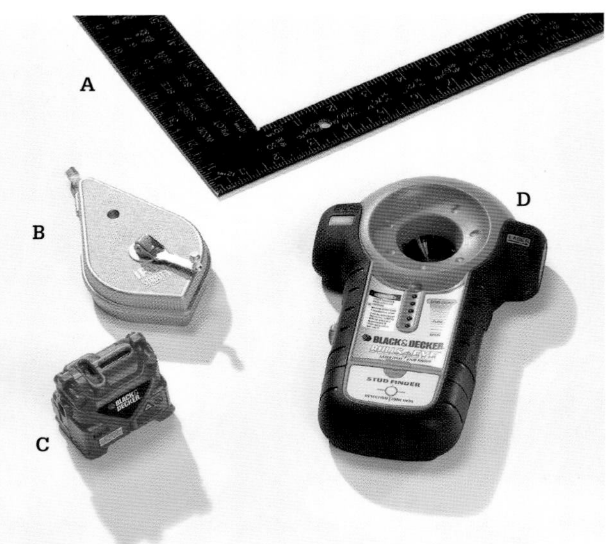

Other useful tools include: a carpenter's square (A), a chalkline (B), a laser level (C), and a combination laser level and stud finder (D).

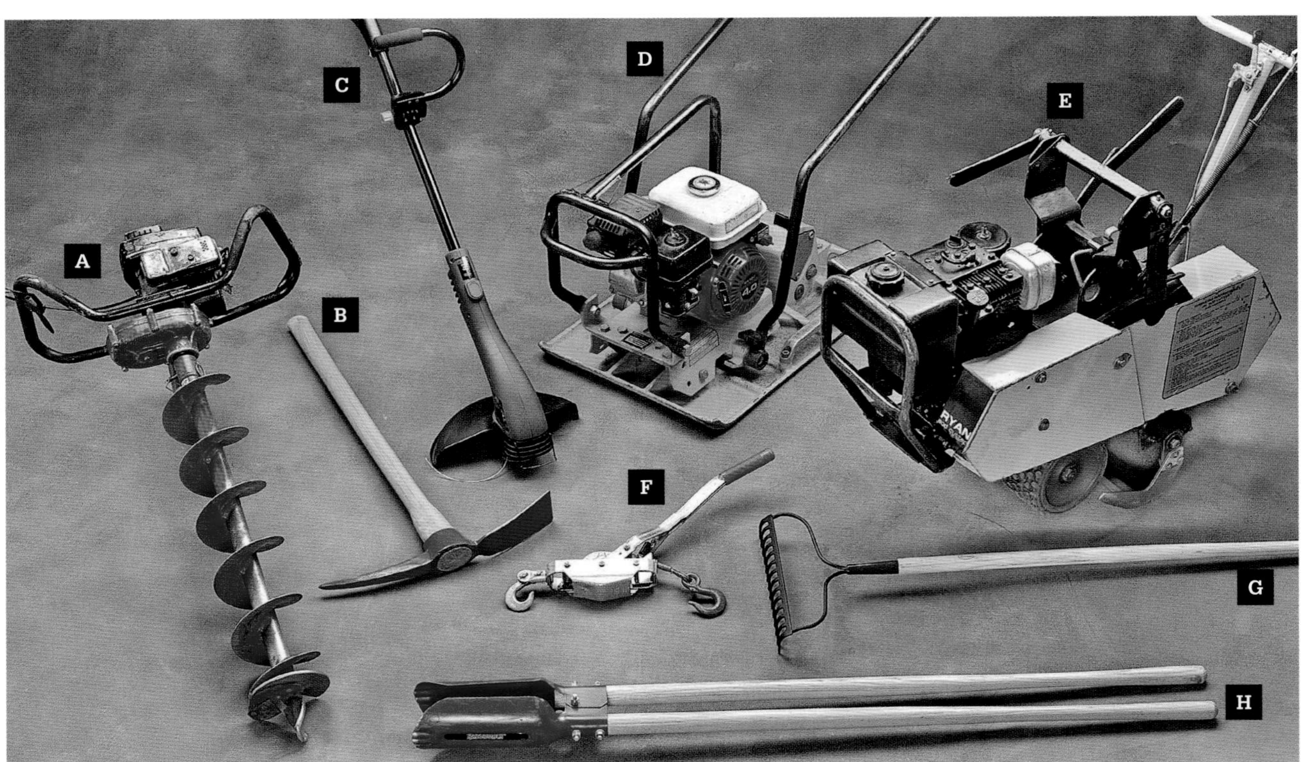

Landscaping tools for preparing sites for concrete projects include: power auger (A) for digging holes for posts or poles; pick (B) for excavating hard or rocky soil; weed trimmer (C) for removing brush and weeds before digging; power tamper (D) and power sod cutter (E) for driveway and other large-scale site preparation; come-along (F) for moving large rocks and other heavy objects; garden rake (G) for moving small amounts of soil and debris; and posthole digger (H) for when you have just a few holes to dig.

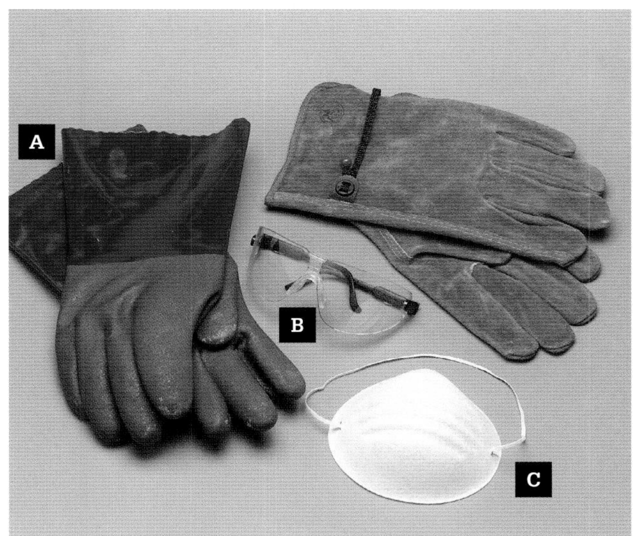

Safety tools and equipment include: gloves (A), safety glasses (B), particle masks (C), and tall rubber boots (not shown). Wear protective gear when handling dry or mixed concrete. These mixes are very alkaline and can burn eyes and skin. When mixing dry bagged concrete, a half-mask respirator is good insurance.

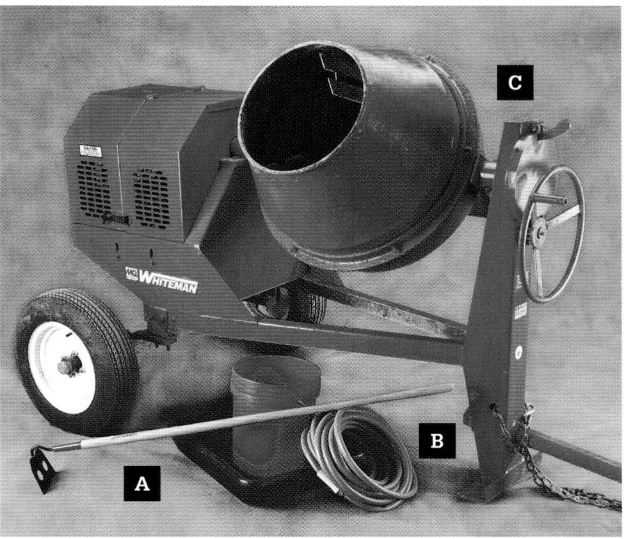

Mixing and pouring tools include: masonry hoe and mortar box (A) for mixing small amounts of concrete; garden hose and bucket (B) for delivering and measuring water; and power mixer (C) for mixing medium-sized (between 2 and 4 cu. ft.) loads of concrete.

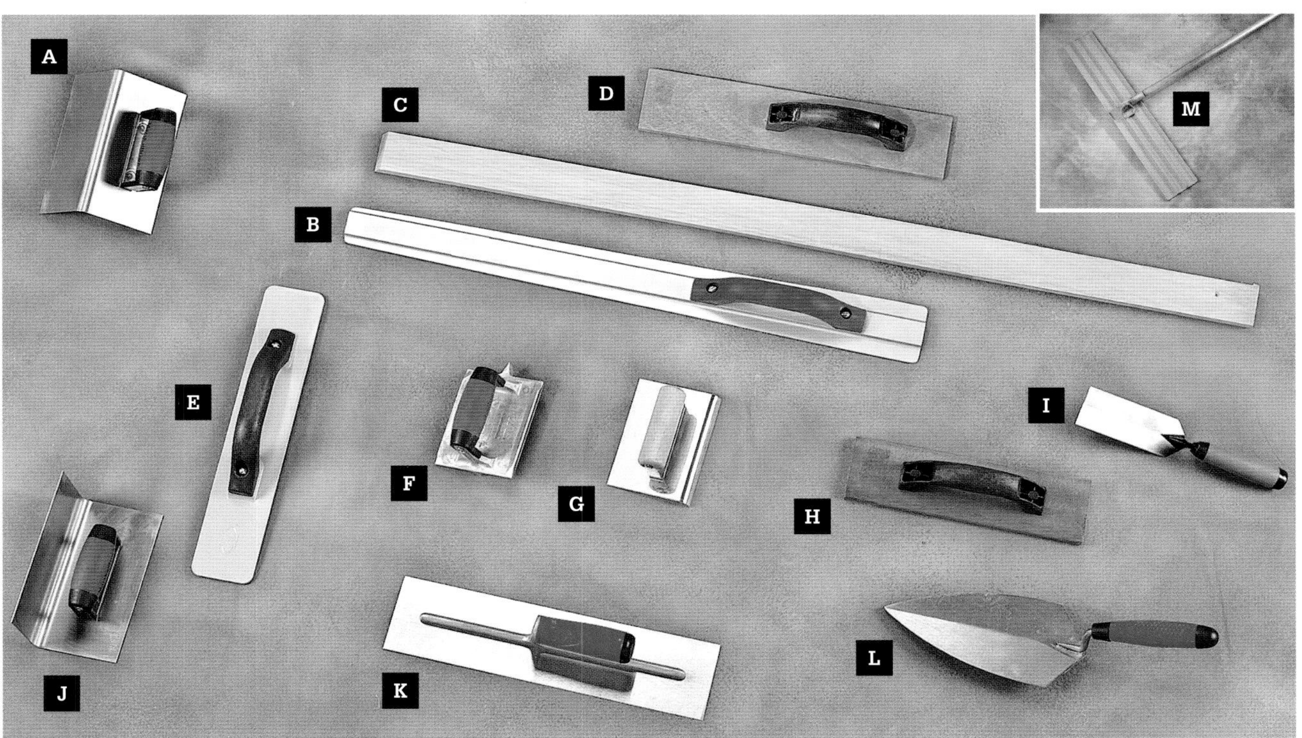

Finishing tools include: outside corner tool (A); aluminum darby (B) for smoothing screeded concrete; screed board (C) for striking off placed concrete; long wood float (D); trowel (E); groover (F) for forming control joints; edger (G) for shaping and forming edges; standard-length wood float (H); square-end trowel (I) for finishing; inside corner tool (J); a steel trowel (K); mason's trowel (L); and long-handled bull float (M) for smoothing large slabs.

Bagged concrete mix comes in many formulations. The selection you're likely to encounter varies by region and by time of year, but the basic products most home centers stock include: all-purpose concrete (A, C) for posts, footings, and slabs; sand mix (B) for topping and casting; Portland cement (D) for mixing with aggregate, sand, and water to make your own concrete; high/early concrete (E) for driveways and other projects that demand greater shock and crack resistance; fast-setting concrete (F) for setting posts and making repairs; specialty blends for specific purposes, such as countertop mix (G,) which comes premixed with polyester fibers and additives that make it suitable for countertops.

Liquid concrete products are either added to the concrete mix while blending or applied after the concrete sets up. Concrete sealer (A) is added to the concrete while liquid for even coloring that goes all the way through the material. Dry pigments also may be added to the wet mixture or scattered onto the surface of concrete slabs during the troweling stage. Bonding additive (B) is applied (usually by roller or sprayer) to protect the concrete from moisture penetration and to give it a deeper, wet-look finish. The product seen here also encourages proper curing when applied immediately after the concrete dries. Concrete colorant (C) is added to dry mix instead of water to make the concrete more elastic and to help it grab onto old concrete surfaces. Also called acrylic fortifier or latex bonding agent, it can be painted onto the old concrete surfaces, as well as the new, to increase concrete adhesion. Stucco and mortar color (D) can be added to finish coat stucco, mason mix, surface-bonding cement, and heavy-duty masonry coating. It is often premixed with water.

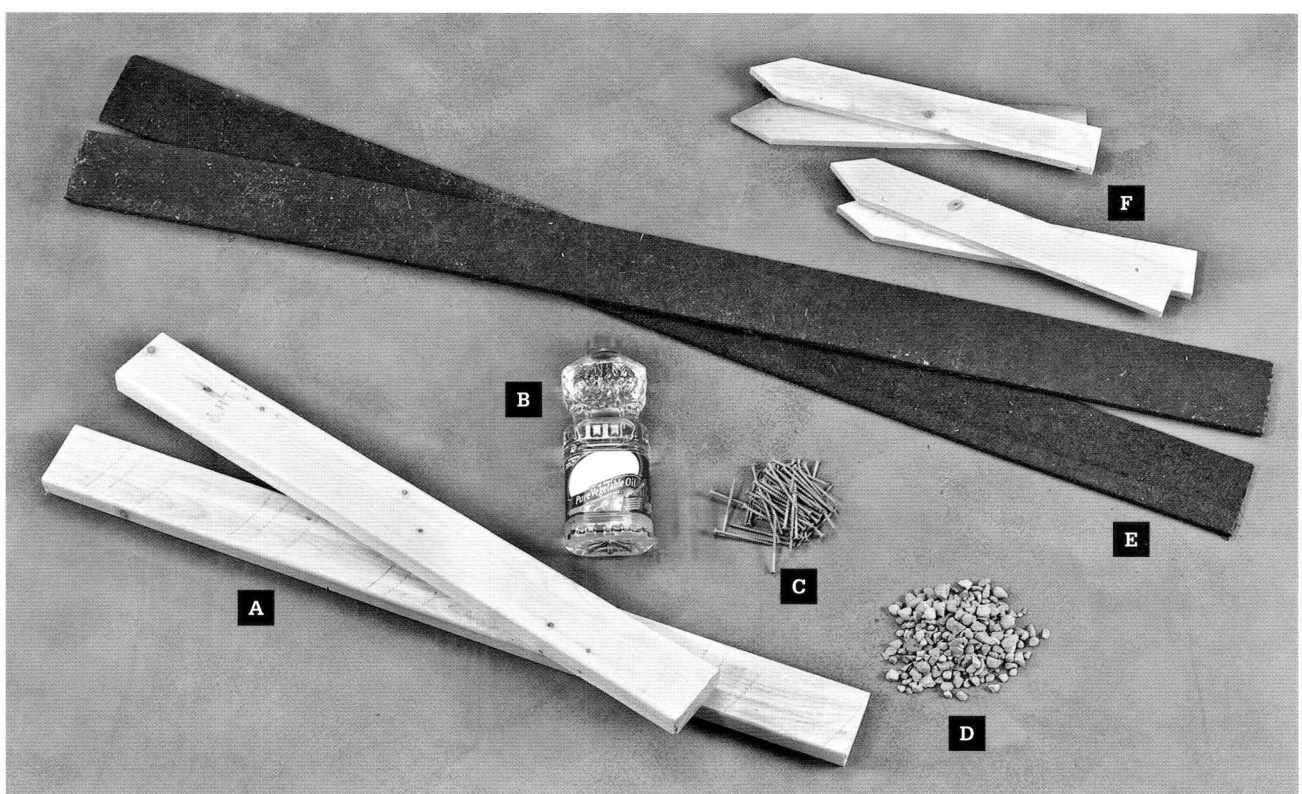

Materials for sub-bases and forms include: lumber (A) and 3" screws (C) for building forms, vegetable oil (B) or a commercial release agent to make it easier to remove the forms, compactable gravel (D) to improve drainage beneath the poured concrete structure, asphalt-impregnated fiberboard (E) to keep concrete from bonding with adjoining structures, and stakes (F) for holding the forms in place.

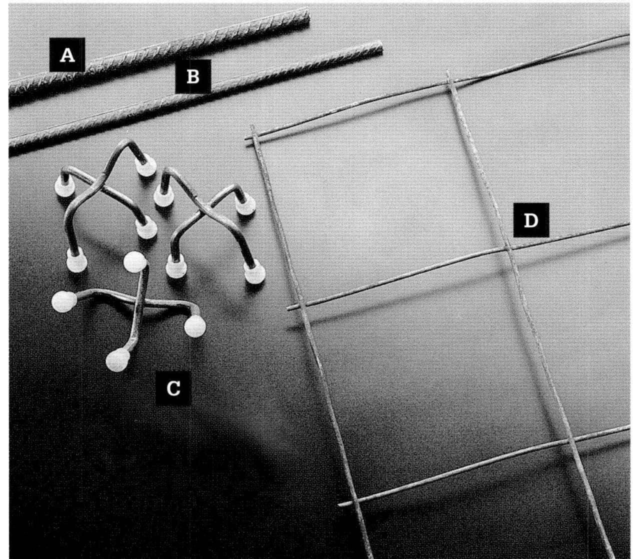

Reinforcement materials: Metal rebar (A, B), available in sizes ranging from #2 (¼" diameter) to #5 (⅝" diameter) reinforce concrete slabs, like sidewalks, and masonry walls; for broad surfaces, like patios, bolsters (C) support rebar and wire mesh; wire mesh (D) (sometimes called remesh) is most common in 6 × 6" grids.

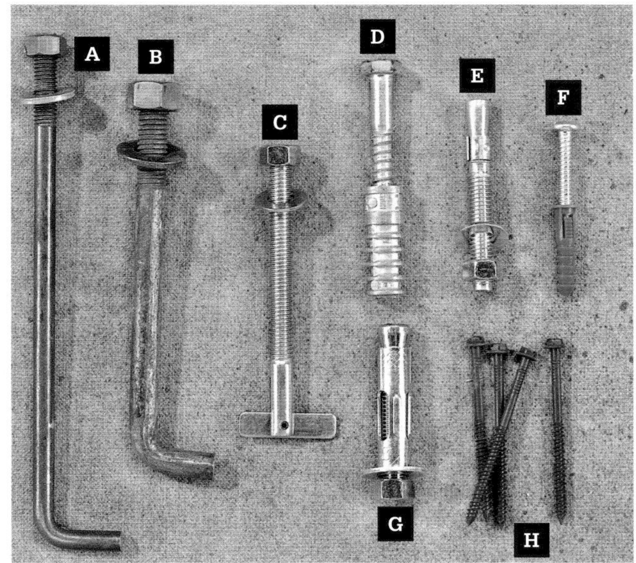

Masonry fasteners allow you to mount objects to concrete and other masonry surfaces. It is most effective to embed the fasteners in fresh concrete so that it cures around the hardware. Examples include: J-bolt with nuts and washers (A, B); removable T-anchor (C); metal sleeve anchor (D); compression sleeves (E, G); light-duty plastic anchor sleeve (F); self-tapping coated steel screws (H).

Planning Concrete Projects

There are two basic stages to planning and designing a concrete project. First is the idea-gathering phase: Employ a variety of sources to help answer questions such as "What qualities will make your project attractive, practical, and durable?" Consider these goals as you plan. Second, apply the basic standards of construction to create a sound plan that complies with local building codes. The projects in this book outline those processes.

Begin by watching for good ideas wherever you go. As you walk through your neighborhood, look for similar projects and observe detail and nuance. Once you've settled on a plan, test the layout by using a rope or hose to outline the proposed project areas. Remember that successful structures take into account size and scale, location, slope and drainage, reinforcement, material selection, and appearance. It's also advisable to take your own level of skills and experience into account, especially if you haven't worked with poured concrete before.

Finally, develop plan drawings for your project. If it's a simple project, quick sketches may be adequate. If permits will be necessary for your project, the building inspector is likely to request detailed plan drawings. (Always check with the local building department early in the planning process.) Either way, drawings help you recognize and avoid or deal with challenges inherent to your project.

Make scaled plan drawings using graph paper and drafting tools. Plan drawings help you eliminate design flaws and accurately estimate material requirements.

Test project layouts before committing to your ideas. Use a rope or hose to outline the area, placing spacers where necessary to maintain accurate, even dimensions.

Estimating Concrete

Plan drawings are especially valuable when the time comes to estimate materials. With your drawing in hand, calculate the width and length of the project in feet, and then multiply the dimensions to get the square footage. Measure the thickness in feet (four inches thick equals one-third of a foot), and then multiply the square footage times the thickness to the get the cubic footage. For example, 1 foot × 3 feet × ⅓ foot = 1 cubic foot. Twenty-seven cubic feet equals one cubic yard.

Coverage rates for poured concrete are determined by the thickness of the slab. The same volume of concrete will yield less surface area if the thickness of the slab is increased.

Unless you need to improve the drainage or the soil stability of your building site, the only other raw material you'll need for most pours is sub-base material, such as compactable gravel. Gravel is normally sold by the cubic yard. For a slab with a four- to six-inch–thick sub-base, one cubic yard of gravel will cover 60 to 70 square feet of surface.

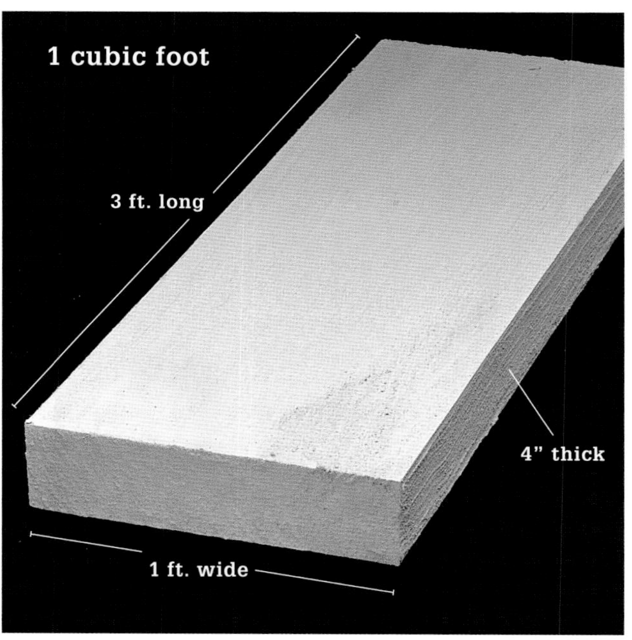

Estimate the materials required based on the dimensions of your project.

Concrete Coverage ▸

Volume	Slab thickness	Surface area
1 cu. yd.	2"	160 sq. ft.
1 cu. yd.	3"	110 sq. ft.
1 cu. yd.	4"	80 sq. ft.
1 cu. yd.	5"	65 sq. ft.
1 cu. yd.	6"	55 sq. ft.
1 cu. yd.	8"	40 sq. ft.

This chart shows the relationship between slab thickness, surface area, and volume.

Estimating Concrete for Tube Format ▸

Footing depth	Number of 60-lb. bags for each size (diameter of tube)			
	6"	8"	10"	12"
1 ft.	1	1	2	2
2 ft.	1	2	3	4
3 ft.	2	3	4	6
4 ft.	2	4	5	7

Footing depth	Number of 80-lb. bags for each size (diameter of tube)			
	6"	8"	10"	12"
1 ft.	1	1	1	2
2 ft.	1	2	2	3
3 ft.	2	3	3	4
4 ft.	2	3	4	6

This chart shows the number of bags of concrete mix a footing requires.

Ordering Concrete

For concrete pours where more than one cubic yard of concrete is required, it usually makes sense to find an alternative to mixing the concrete in small batches. In most cases, this means ordering premixed concrete (called readymix) and having it delivered in a concrete truck to your jobsite. This will increase the concrete cost considerably, especially for smaller pours of one to two cubic yards (most concrete mixing trucks can carry up to six cubic yards, but this is more than a DIYer typically can deal with in one shot). The cost of the material itself is comparable to buying bagged concrete and mixing it yourself, but the minimum delivery charge often exceeds the cost of the material and drives up the total cost.

In addition to saving all the time and energy that is required for mixing concrete by hand, having readymix delivered assures that you will be getting quality material that is customized for your project. When you contact your local, reputable concrete supplier, the first question he or she will ask is "What are you building?" The answer to this (driveway, foundation wall, patio, etc.) allows the supplier to design a mixture that has just the right ratios of ingredients and admixtures that you need. The concrete is further customized on the day of the pour as a good supplier will factor in the weather conditions and the distance from the plant to your house when making the final blend of concrete. The result is a custom blend that will almost certainly outperform anything you can whip up by just adding water. But of course, the tradeoff is higher cost and the inability to control the timing of the pour.

For larger pours there really is no good alternative to having readymix concrete delivered. Once you factor in the delivery charges, it costs a good deal more than mixing your own from bags, but the quality of the material is consistently superior.

Are You Ready for Delivery?

1. Are all concrete forms built and staked?
2. Has reinforcement (rebar or remesh) been placed inside the forms as required?
3. Has clean fill been placed in form as required (mostly used with steps)?
4. Have forms been treated with a release agent?
5. Have any required permits to block the sidewalk or street been obtained? Ask your concrete supplier for more information.
6. Has a path from the truck parking spot to the project been cleared?
7. Have you consulted with the readymix supplier to determine if wheelbarrows will be necessary?
8. Do you have at least two heavy-duty wheelbarrows (and strong helpers to run them) on site?
9. Are trowels, floats, and screeds clean and ready to go?
10. Do you have a concrete vibrator rented and on site, if required?
11. Have you constructed a ramp for wheelbarrows to scale forms and get inside project area (where required)?
12. Have you confirmed the delivery time, amounts, and type of concrete with the readymix supplier?
13. **VERY IMPORTANT: Have your forms been inspected and approved by your local building department?**

How Much Does Concrete Cost?

The actual costs of buying bagged concrete or having concrete mixed and delivered to your site can be hard to predict. Like gasoline or plywood, concrete is a commodity and is subject to fairly wild price fluctuations as market conditions change. The costs also vary quite a bit regionally and seasonally. But for the sake of comparison, here are the results of some recent price shopping (quotes obtained on the same day in the same city), along with some notes on each method.

BAGGED CONCRETE MIXED BY HAND:

Cost of materials: $3.50 per 80 lb. bag (general-purpose concrete)
80 lb. bags per cubic yard: 41
Cost of 1 cubic yard: $143.50

Notes:

- One person mixing concrete in a wheelbarrow or mortar box and working hard can expect to mix one 80-pound bag in about five minutes (or close to three and a half hours per cubic yard if mixing nonstop). If you are pouring footings or if your project is broken up into smaller sections, you may be able to mix the concrete by hand, especially if you halve the mixing time by recruiting a helper and an extra wheelbarrow. But for large slabs and walls, hand mixing is too slow. You will not get consistency of water content, which can lead to cracking along seams between batches. Tooling also works better if you can finish the whole project at one time.
- Electric or gas-powered mixers do save a lot of time. A larger gas-powered mixer that can handle three 80-pound bags will yield two cubic feet of mixed concrete in about five minutes.
- General-purpose concrete (usually around 3,500 psi) is relatively inexpensive in bagged form, but if you use special blends the cost will go upward very fast. High-strength and fast-setting concrete cost about twice as much as general-purpose, and special purpose mixtures, such as countertop, concrete can cost five times as much.

READYMIX CONCRETE IN DIY TRAILER

Trailer rental: Waived with purchase
Cost of ½ cubic yard: $117
Cost of 1 cubic yard: $155

Notes:

- If you will be pulling the trailer with a vehicle smaller than a full-size pickup, you probably won't be allowed to haul more than one half cubic yard of mixed concrete at a time. If this means you'll be making multiple trips, be sure to place your entire order at once so you will be charged the discounted rate for larger orders.
- Freshly mixed concrete is bounced and vibrated in the trailer as it rolls along over the road. A concrete mixer on board a concrete truck turns continually, so road vibration does not create a problem. But in a trailer, separation of the liquid and the heavier solids will occur during transport, weakening the concrete. If you need to haul the concrete more than 10 or 15 miles, a trailer is not recommended.

READYMIX CONCRETE (DELIVERED)

Delivery Charge: $100 to $150 on average
Cost of 1 cubic yard: $90
Total cost: $240 (per-yard cost drops off dramatically for larger orders)

Notes:

- Operator charges: Concrete supplier may add additional charge for driver time beyond a minimum allowance of free time (5 or 10 minutes per yard is typical)
- Customizing your concrete with admixtures such as air entrainers and accelerants usually doesn't have a major impact on the cost of ready mix, as it does with bagged mixtures. Adding tint to the mixture can result in a surcharge of 50 percent or more, however.
- Some drivers are happy to jump in and lend a hand with the work. Others prefer to stay in their climate-controlled cabs. If you get lucky, that's great—but don't plan on it.

Mixing Concrete

When mixing concrete on site, purchase bags of dry premixed concrete and simply add water. Follow the instructions carefully and take note of exactly how much water you add so the concrete will be uniform from one batch to the next. Never mix less than a full bag, however, since key ingredients may have settled to the bottom.

For smaller projects, mix the concrete in a wheelbarrow or mortar box. For larger projects, rent or buy a power mixer. Be aware that most power mixers should not be filled more than half full.

When mixing concrete, the more water you add, the weaker the concrete will become. For example, if you need "slippery" concrete to get into the corners of a form, add a latex bonding agent or acrylic fortifier instead of water. Mix the concrete only until all of the dry ingredients are moistened; don't overwork it.

Tools & Materials ▸

- Power mixer or mortar box
- Wheelbarrow
- Masonry hoe
- 5-gal. Bucket
- Safety glasses
- Gloves
- Particle mask
- Hammer
- 2 × 4 lumber
- Bagged concrete mix
- Clean, fresh water
- Mason's trowel

Too dry

Too wet

Correct

A good mixture is crucial to any successful concrete project. Properly mixed concrete is damp enough to form in your hand when you squeeze and dry enough to hold its shape. If the mixture is too dry, the aggregate will be difficult to work and will not smooth out easily to produce an even, finished appearance. A wet mixture will slide off the trowel and may cause cracking and other defects in the finished surface.

How to Mix Concrete

Empty premixed concrete bags into a mortar box or wheelbarrow. Form a hollow in the mound of dry mix, and then pour water into the hollow. Start with ¾ of the estimated water amount per 80-lb. bag.

Work the material with a hoe, continuing to add water until a pancake batter consistency is achieved. Clear out any dry pockets from the corners. Do not overwork the mix. Also, keep track of how much water you use in the first batch so you will have a reliable guideline for subsequent batches.

Using a Power Mixer

Fill a bucket with ¾ gal. of water for each 80-lb. bag of concrete you will use in the batch for most power mixers, 3 bags is a workable amount. Pour in half the water. Before you start power-mixing, carefully review the operating instructions for the mixer.

Add all of the dry ingredients, and then mix for 1 minute. Pour in water as needed until the proper consistency is achieved and mix for 3 to 5 minutes. Pivot the mixing drum to empty the concrete into a wheelbarrow. Rinse out the drum immediately.

Placing Concrete

Placing concrete involves pouring it into forms, and then leveling and smoothing it with special masonry tools. Once the surface is smooth and level, control joints are cut and the edges are rounded. Special attention to detail in these steps will result in a professional appearance. *Note: If you plan to add a special finish, read "Curing & Finishing Concrete" (page 34) before you begin your project. Be sure to apply a release agent before pouring the concrete.*

Tools & Materials ▸

- Wheelbarrow
- Shovel
- Lumber
- 3" screws
- Rebar
- Sand
- Bucket
- Gloves
- Safety glasses
- Masonry hoe
- Mixed concrete
- Spade
- Hammer
- Concrete vibrator
- Float
- Groover
- Edger

Moving concrete from the source to the destination represents much of the work in many concrete pours. Make sure to plan ahead to create access for wheelbarrows or ready-mix trucks, and get all of the help you can round up for the actual pour.

Do not overload your wheelbarrow. Experiment with sand or dry mix to find a comfortable, controllable volume. This also helps you get a feel for how many wheelbarrow loads it will take to complete your project.

Lay planks over the forms to make a ramp for the wheelbarrow. Avoid disturbing the building site by using ramp supports. Make sure you have a flat, stable surface between the concrete source and the forms.

How to Place Concrete

Load the wheelbarrow with fresh concrete. Clear a path from the source to the site. Always load wheelbarrows from the front; loading from the side can cause tipping.

Pour concrete in evenly spaced loads. Start at the farthest point from the concrete source, and work your way back. Pour so concrete is a few inches above the tops of the forms. If you're using wood scrap as shims to support a ramp end, be sure and fasten them together with screws or nails.

Continue placing concrete, working away from your starting point. Do not pour more concrete than you can tool at one time. Monitor the concrete surface to make sure it does not harden too much before you can start tooling.

Distribute concrete evenly in the project area using a masonry hoe. Work the concrete with a hoe until it is fairly flat, and the surface is slightly above the top of the forms. Remove excess concrete from the project area with a shovel. Avoid overworking the concrete and take care not to disturb reinforcement.

(continued)

Work a spade blade between the inside edges of the forms and the concrete to remove trapped air bubbles that can weaken the material. This will help settle the concrete.

Hammer the sides of forms to settle the concrete. This action draws finer aggregates in the concrete against the forms, creating a smoother surface on the sides. This is especially important when building steps. For larger pours, rent a concrete vibrator for this job (inset).

Remove excess concrete with a screed board—a straight piece of 2 × 4 lumber or angle iron long enough to rest on opposite forms. Move the board in a sawing motion, and keep it flat as you work. If screeding leaves valleys in the surface, add fresh concrete in the low areas and screed them to level.

Float the surface in an arching motion once bleed water disappears. Float with the leading edge of the tool tipped up, and stop floating as soon as the surface is smooth so you do not overwork the concrete.

9

Draw a groover tool at joint locations after bleed water has dried. Use a straight 2 × 4 as a guide. You may need to make several passes to create a smooth control joint. Avoid digging into the concrete surface with the flat edges of the tool. Smooth out tool marks with a float once the joint is cut.

Shape concrete with an edging tool, smoothing between the forms and concrete to create a finished appearance. Make several passes if necessary and use a float to smooth out any marks left by the groover or edger. *Note: Cutting a radiused profile in the edges also prevents concrete from cracking and chipping.*

Understanding Bleed Water ▸

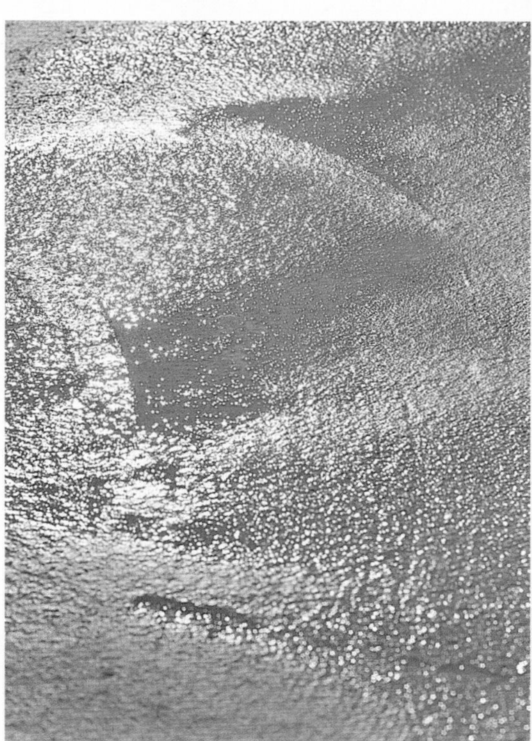

Timing is key to an attractive concrete finish. When concrete is poured, the heavy materials gradually sink, leaving a thin layer of water—known as bleed water—on the surface. To achieve an attractive finish, it's important to let bleed water evaporate before proceeding with other steps. Follow these rules to avoid problems:

- Settle and screed the concrete and add control joints (steps 9 through 11) immediately after pouring and before bleed water appears. Otherwise, crazing, spalling, and other flaws are likely.
- Let bleed water dry before floating or edging. Concrete should be hard enough that foot pressure leaves no more than a ¼"-deep impression.
- Do not overfloat the concrete; it may cause bleed water to reappear. Stop floating if a sheen appears and resume when it is gone.

Note: Bleed water may not appear with air-entrained concrete, which is used in regions where temperatures often fall below freezing.

Curing & Finishing Concrete

Concrete that is exposed, such as for patios and sidewalks, requires a good finish for project success. Creating the final finish may be as simple as troweling the surface and letting it dry. Or, you may choose to do something a little more decorative, such as an acid stain, a broomed antiskid surface, or exposed aggregate. Some of the fancier finishes you can do with concrete are covered in the Decorative Masonry Finishing chapter. Because it is done when the concrete is still wet, we've included some information on exposed aggregate finishes here.

Once the fresh concrete has been edged and the control joints have been cut, it needs to dry for a period of time before any surface finishing, such as brooming or exposing aggregate, can be done. After that, it should dry overnight before any forms are removed. Finally, it should cure for three to seven days, or even longer depending on the type of concrete, the conditions, and the nature of the project. Traditionally, concrete is covered with burlap or sheet plastic for the drying and curing phases, and the surface is dampened a couple of times a day to slow down the process. Concrete that dries too fast can crack. However, most professionals today have recognized that covering the concrete often causes more problems than it prevents. So they are less likely to cover the concrete, preferring instead to treat it with a curing or sealing agent once it sets up, or simply to let it dry naturally in the open air.

Tools & Materials ▸

- Broom
- Wheelbarrow
- Shovel
- Magnesium float
- Groover
- Edger
- Hose
- Coarse brush
- Plastic sheeting
- Aggregate
- Water
- 2 × 4 lumber
- Screed board
- Muriatic acid
- Work gloves and eye protection

Tip for a Broomed Finish ▸

For a nonslip finish, tool the concrete and then draw a clean stiff-bristle broom across the surface once the concrete is thumbprint hard. Wait until concrete is firm to the touch to achieve a finer texture and a more weather-resistant surface. Make sure all strokes are made in the same direction and avoid overlapping.

How to Create an Exposed Aggregate Finish

Exposed aggregate finishes provide a rugged, nonskid surface that is resistant to heavy traffic and extreme weather. This effect helps the concrete surface blend into a landscape featuring natural stone. Chose from a wide variety of decorative aggregate to achieve different finishes.

Place the concrete. After smoothing the surface with a screed board, let any bleed water disappear; then spread clean, washed aggregate evenly with a shovel or by hand. Spread smaller aggregate (up to 1" in dia.) in a single layer; for larger aggregate, maintain a separation between stones that is roughly equal to the size of one stone.

Pat the aggregate down with the screed board, and then float the surface with a magnesium float until a thin layer of concrete covers the stones. Do not overfloat. If bleed water appears, stop floating and let it dry before completing the step. If you are seeding a large area, cover it with plastic to keep the concrete from hardening too quickly.

Cut control joints and tool the edges. Let concrete set for 30 to 60 minutes, and then mist a section of the surface and scrub with a brush to remove the concrete covering the aggregate. If brushing dislodges some of the stones, reset them and try again later. When you can scrub without dislodging stones, mist and scrub the entire surface to expose the aggregate. Rinse clean. Do not let the concrete dry too long, or it will be difficult to scrub off.

Rinse the concrete surface with a hose after it has cured for one week and covering is removed. If a residue remains, try scrubbing it clean. If scrubbing is ineffective, wash the surface with a muriatic acid solution, and then rinse immediately and thoroughly with water.

Preparing a Project Site

The first stage of any poured concrete project is to prepare the project site. The basic steps include the following:

1. Lay out the project using stakes and strings.
2. Clear the project area and remove the sod.
3. Excavate the site to allow for a sub-base and footings (if necessary) and concrete.
4. Lay a sub-base for drainage and stability and pour footings (if necessary).
5. Build and install reinforced wood forms.

Proper site preparation depends on the project and site. Plan on a sub-base of compactible gravel. Some projects require footings that extend past the frost line, while others, such as sidewalks, do not. Consult your local building inspector about the specific requirements of your project.

If your yard slopes more than one inch per foot, you may need to add or remove soil to level the surface. A landscape engineer or building inspector can advise you on how to prepare a sloping project site.

Tools & Materials ▸

Rope
Carpenter's square
Hand maul
Tape measure
Mason's string
Line level
Spade
Sod cutter
Straightedge
Level
Wheelbarrow
Shovel
Hand tamper
Circular saw
Drill
Lumber (2 × 4, 1 × 4)
3" Screws
Compactable gravel
Vegetable oil or commercial release agent
Asphalt-impregnated fiber board
Wood stakes
Work gloves
Common nails

Good site preparation is one of the keys to a successful project. Patience and attention to detail when excavating, building forms, and establishing a sub-base help ensure that your finished project is level and stable and will last for many years.

How to Prepare the Site

Measure the slope of the building site to determine if you need to do grading work before you start your project. First, drive stakes at each end of the project area. Attach a mason's string between the stakes and use a line level to set it at level. At each stake, measure from the string to the ground. The difference between the measurements (in inches) divided by the distance between stakes (in feet) will give you the slope (in inches per foot). If the slope is greater than 1" per foot, you may need to regrade the site.

Dig a test hole to the planned depth so you can evaluate the soil conditions and get a better idea of how easy the excavation will be. Sandy or loose soil may require amending; consult a landscape engineer.

Add a compactable gravel sub-base to provide a level, stable foundation for the concrete. For most building projects, pour a layer of compactable gravel about 4 to 6" thick, and use a tamper to compress it to 4".

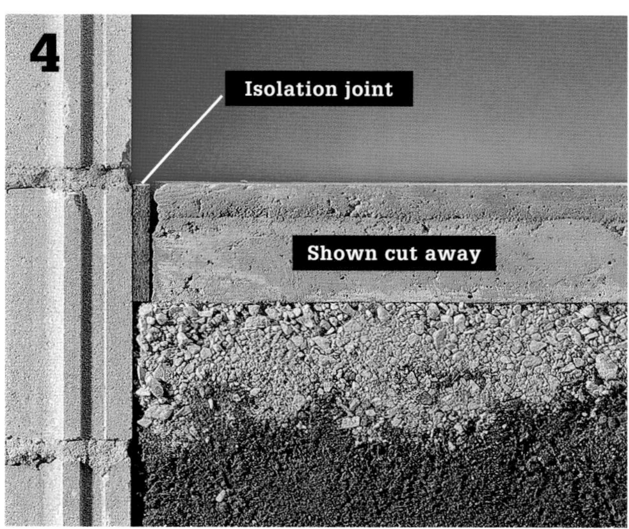

When pouring concrete next to structures, glue a ½"-thick piece of asphalt-impregnated fiber board to the adjoining structure to keep the concrete from bonding with the structure. The board creates an isolation joint, allowing the structures to move independently and minimizing the risk of damage.

How to Lay Out & Excavate the Site

Lay out a rough project outline with a rope or hose. Use a carpenter's square to set perpendicular lines. To create the actual layout, begin by driving wood stakes near each corner of the rough layout. The goal is to arrange the stakes so they are outside the actual project area, but in alignment with the borders of the project. Where possible, use two stakes set back 1 ft. from each corner, so strings intersect to mark each corner (below). *Note: In projects built next to permanent structures, the structure will define one project side.*

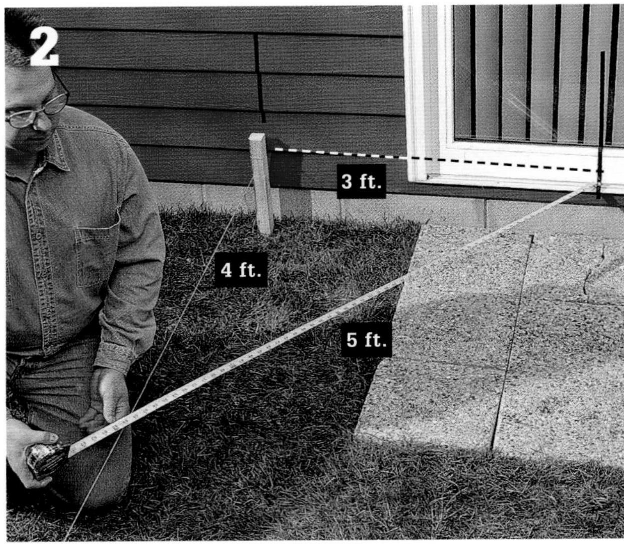

Connect the stakes with mason's strings. The strings should follow the actual project outlines. To make sure the strings are square, use the 3-4-5 triangle method: Measure and mark points 3 ft. out from one corner along one string and 4 ft. out along the intersecting string at the corner. Measure between the points, and adjust the positions of the strings until the distance between the points is exactly 5 ft. A helper will make this easier.

Reset the stakes, if necessary, to conform to the positions of the squared strings. Check all corners with the 3-4-5 method, and adjust until the entire project area is exactly square. This can be a lengthy process with plenty of trial and error, but it is very important to the success of the project, especially if you plan to build on the concrete surface.

Attach a line level to one of the mason's strings to use as a reference. Adjust the string up or down as necessary until it is level. Adjust the other strings until they are level, making sure that intersecting strings contact one another. This ensures that they are all at the same height relative to ground level.

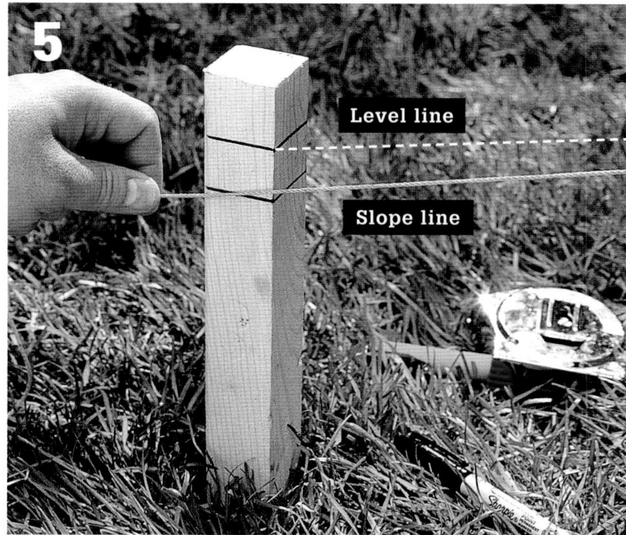

Shift mason's strings on opposite sides of the project downward on their stakes (the lower end should be farther away from the house). Most concrete surfaces should have a slight slope to direct water runoff, especially if the surface is near your house. To create a standard slope of ⅛" per foot, multiply the distance between the stakes on one side (in feet) by ⅛. For example, if the stakes were 10 ft. apart, the result would be 10/8 (1¼"). You would move the strings down 1¼" on the stakes on the low ends.

Start excavating by removing the sod. Use a sod cutter if you wish to reuse the sod elsewhere in your yard (lay the sod as soon as possible). Otherwise, use a square-end spade to cut away sod. Strip off the sod at least 6" beyond the mason's strings to make room for 2 × 4 forms. You may need to remove the strings temporarily for this step.

Make a story pole as a guide for excavating the site. First, measure down to ground level from the high end of a slope line. Add 7½" to that distance (4" for the sub-base material and 3½" for the concrete if you are using 2 × 4 forms). Mark the total distance on the story pole, measuring from one end. Remove soil from the site with a spade. Use the story pole to make sure the bottom of the site is consistent (the same distance from the slope line at all points) as you dig. Check points at the center of the site using a straightedge and a level placed on top of the soil.

Lay a sub-base for the project (unless your project requires a frost footing). Pour a 5"-thick layer of compactable gravel in the project site, and tamp until the gravel is even and compressed to 4" in depth. *Note: The sub-base should extend at least 6" beyond the project outline.*

How to Build & Install Wood Forms

A form is a frame, usually made from 2 × 4 lumber, laid around a project site to contain poured concrete and establish its thickness. Cut 2 × 4s to create a frame with inside dimensions equal to the total size of the project.

Use the mason's strings that outline the project as a reference for setting form boards in place. Starting with the longest form board, position the boards so the inside edges are directly below the strings.

Cut several pieces of 2 × 4 at least 12" long to use as stakes. Trim one end of each stake to a sharp point. Drive the stakes at 3-ft. intervals at the outside edges of the form boards, positioned to support any joints in the form boards.

Drive 3" deck screws through the stakes and into the form board on one side. Set a level so it spans the staked side of the form and the opposite form board, and use the level as a guide as you stake the second form board so it is level with the first. For large projects, use the mason's strings as the primary guide for setting the height of all form boards.

Once the forms are staked and leveled, drive 3" deck screws at the corners. Coat the insides of the forms with vegetable oil or a commercial release agent so concrete won't bond to them. *Tip: Tack nails to the outsides of the forms to mark locations for control joints at intervals roughly 1½ times the slab's width (but no more than 30 times its thickness).*

Variations for Building Forms ▸

Straight forms made from dimensional lumber do not work for every concrete project. Use ¾"-thick plywood for building taller forms for projects such as concrete steps or walls. Use the earth as a form when building footings for poured concrete building projects. Use standard wood forms for the tops of footings for building with brick or block when the footing will be visible. Create curves with hardboard, siding or other thin, flexible sheet stock attached at the inside corners of a form frame.

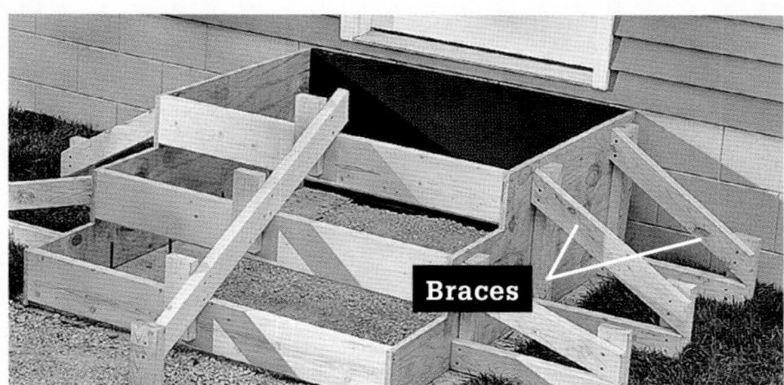

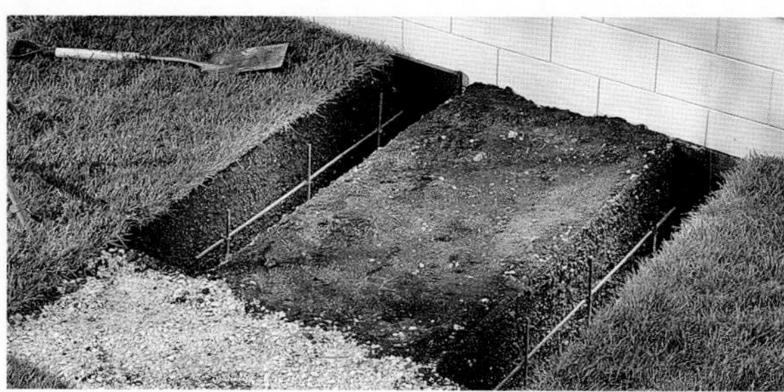

Tips for Working with Rebar & Remesh ▸

Cut rebar with a reciprocating saw that is equipped with a metal-cutting blade (cutting rebar with a hacksaw can take 5 to 10 minutes per cut). Use bolt cutters to cut wire mesh.

Overlap joints in rebar by at least 12", and then bind the ends together with heavy-gauge wire. Overlap seams in wire mesh reinforcement by 12".

Leave at least 1" of clearance between the forms and the edges or ends of metal reinforcement. Use bolsters or small chunks of concrete to raise remesh reinforcement off the sub-base, but make sure it is at least 2" below the tops of the forms.

Setting Concrete Posts

Even among professional landscapers, you'll find widely differing practices for setting fence posts. Some take the always-overbuild approach and set every post in concrete that extends a foot past the frost line. Others prefer the adjustability and improved drainage you get when setting posts in packed sand or gravel. Some treat the post ends with preservative before setting the posts, others don't bother. The posts may be set all at once, prior to installing the stringers and siding; or, they may be set one-at-a-time in a build-as-you-go approach. Ultimately, the decision of whether to set posts in concrete comes down to weight of the panels, as well as where they fall in the fence layout. At the very least, gate posts and corner posts should be set into concrete.

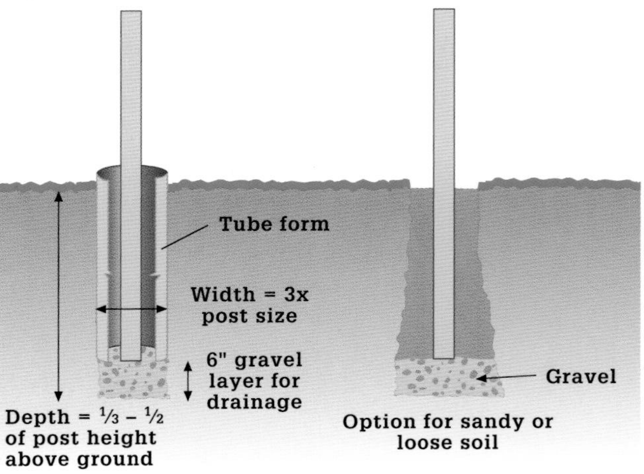

Tools & Materials ▸

- Shovel
- Posthole digger or gas-powered auger
- Digging bar auger
- Hand tamper
- Saw
- Drill
- Maul
- Level
- Trowel
- Gravel
- Concrete mix
- Post lumber (4 × 4)
- 2 × 4 lumber
- Screws
- Wood sealer (for wood posts)
- Plumb bob
- Mason's string
- Paint brush
- Waxed paper
- Gloves and eye protection

When setting posts for fences, one way to ensure that post spacing is exactly correct is to assemble the fence panels and posts prior to digging the postholes. Then, you can lay the fence section or sections in place and mark post locations exactly where the posts will hit. Level and brace the structure and fill in the concrete.

To preserve post ends, treat fence posts with wood preservative before setting them into the ground or into concrete. These chemicals may be applied with a brush or, if you are setting a lot of posts, by dipping them directly into a container. Most wood preservatives are quite toxic so follow all safety precautions and handling recommendations.

How to Set a Post in Concrete

Dig a hole that is three times wider than the post width (or diameter) and as deep as ⅓ the post length, plus 6". Use a posthole digger for most of the digging and a digging bar to dislodge rocks and loosen compacted soil.

Option: Use a gas-powered auger if you are digging several postholes—especially if you need to dig deeper than 20 to 24" (the depth at which posthole diggers become ineffective).

Pour 6" of loose gravel into the bottom of the hole to create drainage. Tamp the gravel, using a hand tamper or wood post.

Set the post in the hole. Attach wood braces to two adjacent faces of the post. Check for plumb, then drive a stake into the ground near the end of each brace, and attach the ends of the braces to the stakes.

Mix all-purpose concrete and pour it into the hole, overfilling it slightly. Tamp the concrete down with the butt end of a 2 × 4 to settle it into the posthole. *TIP: Mask the post temporarily with waxed paper before adding concrete to protect the wood from discoloration and staining. Use a small trowel to smooth the concrete and form a slight crown.*

Concrete Piers

Concrete pier footings support the weight of outdoor structures, such as decks and pergolas. Check local codes to determine the size and depth of pier footings required in your area. In cold climates, footings must be deeper than the soil frost line.

To help protect posts from water damage, each footing should be poured so that it is 2" above ground level. Tube-shaped forms let you extend the footings above ground level.

As an alternative to inserting J-bolts into wet concrete, you can use masonry anchors or install anchor bolts with an epoxy designed for masonry installations.

Before digging, consult local utilities for the location of any underground electrical, telephone, or water lines that might interfere with footings.

Tools & Materials ▸

- Power auger or posthole digger
- Tape measure
- Shovel
- Reciprocating saw or handsaw
- Torpedo level
- Hoe
- Trowel
- Hammer drill
- Shovel
- Utility knife
- Wheelbarrow
- Concrete tube forms
- Concrete mix
- J-bolts or masonry anchors
- Scrap 2 × 4
- Gravel
- Speed square
- Shop vac
- Masking tape
- Epoxy
- Gloves and eye protection

How to Pour Concrete Piers

To install tube forms for concrete piers, dig holes with a posthole digger or power auger. Pour 4 to 6" of gravel in the bottom for drainage, and then cut and insert the tube, leaving about 2" of tube above ground level. Pack soil around tubes to hold them in place.

Slowly pour concrete into the tube, guiding concrete from the wheelbarrow with a shovel. Fill about half of the form. Use a long board to tamp the concrete, filling any air gaps in the footing. Then finish pouring and tamping concrete into the form.

Level the concrete by pulling a wood scrap across the top of the tube form using a sawing motion. Add concrete to any low spots.

Insert a J-bolt into wet cement, lowering it slowly and wiggling it slightly to eliminate any air gaps. Set the J-bolt so ¾ to 1" is exposed above the concrete. Brush away any wet concrete on the bolt threads.

Use a torpedo level to make sure the J-bolt is plumb. If necessary, adjust the bolt and repack concrete. Let concrete dry, and then cut away exposed portion of tube with a utility knife (optional).

How to Install Anchor Bolts in Epoxy

Drill a hole for the threaded rod after the pier has cured at least 48 hours. Locate the bolt locations and drill using a hammer drill and masonry bit sized to match the rod diameter. Use a speed square to vertically align the drill, set the depth gauge so the rod will protrude ¾ to 1" above the pier. After drilling, clean out debris from the hole using a shop vac.

Wrap masking tape around ¾ to 1" of rod for reference. Inject epoxy into hole using the mixing syringe provided by the manufacturer. Use enough epoxy so a small amount is forced from the hole when the rod is fully inserted. Insert the rod immediately; epoxy begins to harden as soon as it is injected. Check the height of the rod, and then allow the epoxy to cure for 16 to 24 hours. If necessary, trim the rod using a reciprocating saw with a metal-cutting blade.

Poured Footings for Freestanding Walls

Footings provide a stable, level base for brick, block, stone, and poured concrete structures. They distribute the weight of the structure evenly, prevent sinking, and keep structures from moving during seasonal freeze-thaw cycles.

The required depth of a footing is usually determined by the frost line, which varies by region. The frost line is the point nearest ground level where the soil does not freeze. In colder climates, it is likely to be 48 inches or deeper. Frost footings (footings designed to keep structures from moving during freezing temperatures) should extend 12 inches below the frost line. Your local building inspector can tell you the frost line depth for your area.

Tips for Planning ▸

- Describe the proposed structure to your local building inspector to find out whether it requires a footing and whether the footing needs reinforcement. In some cases, 8"-thick slab footings can be used as long as the sub-base provides plenty of drainage.
- Keep footings separate from adjoining structures by installing an isolation board (page 37).
- For smaller poured concrete projects, consider pouring the footing and the structure as one unit.
- A multi-wall project such as a barbecue may require a floating footing (page 147).

Footings are required by building code for concrete, stone, brick, and block structures that adjoin other permanent structures or that exceed the height specified by local codes. Frost footings extend 8 to 12" below the frost line. Slab footings, which are typically 8" thick, may be recommended for low, freestanding structures built using mortar or poured concrete. Before starting your project, ask a building inspector about footing recommendations and requirements for your area.

Tools & Materials ▸

Rope
Carpenter's square
Hand maul
Tape measure
Mason's string
Line level
Spade
Shovel
Sod cutter
Straightedge
Level
Wheelbarrow
Hand tamper
Circular saw
Reciprocating saw
Isolation board
Float
Drill
Concrete mix
#3 rebar
16-gauge wire
2 × 4 lumber
3" screws
Compactable gravel
Vegetable oil or release agent
Stakes
Tie-rods
Eye protetction and work gloves

Options for Forming Footings

When you build footings, make them twice as wide as the wall or structure they will support. Footings also should extend at least 12 inches past the ends of the project area. Add two tie rods if you will be pouring concrete over the footing. After the concrete sets up, press 12-inch pieces of rebar six inches into the concrete. The tie-rods will anchor the footing to the structure it supports.

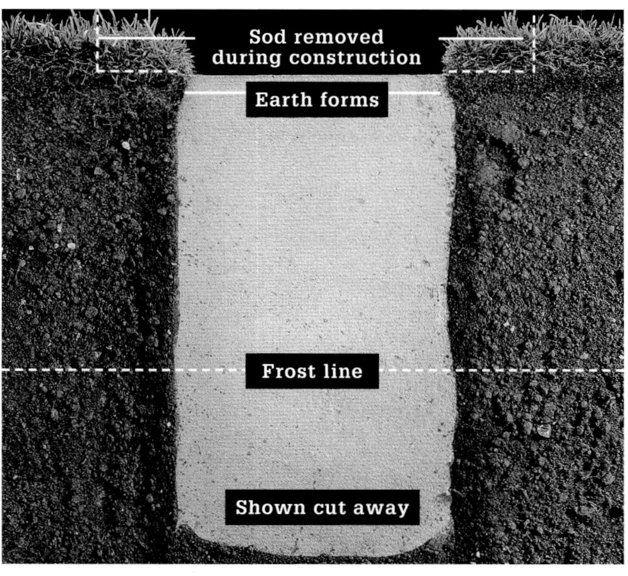

For poured concrete, use the earth as a form. Strip sod from around the project area, and then strike off the concrete with a screed board resting on the earth at the edges of the top of the trench.

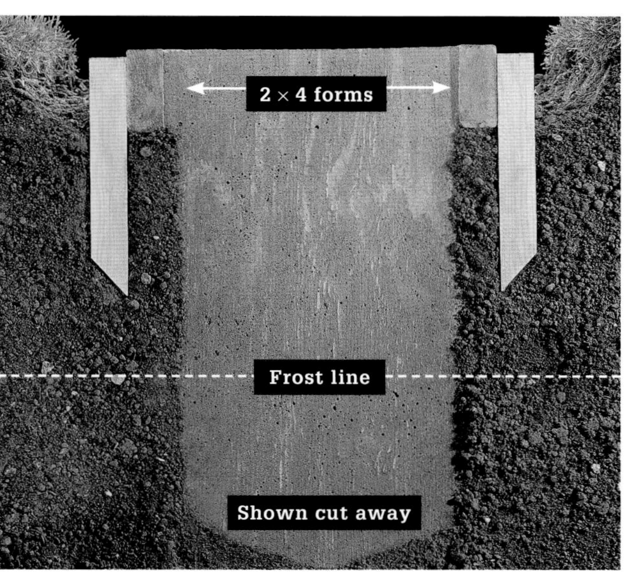

For brick, block, and stone, build level, recessed wood forms. Rest the screed board on the frames when you strike off the concrete to create a flat, even surface for stacking masonry units.

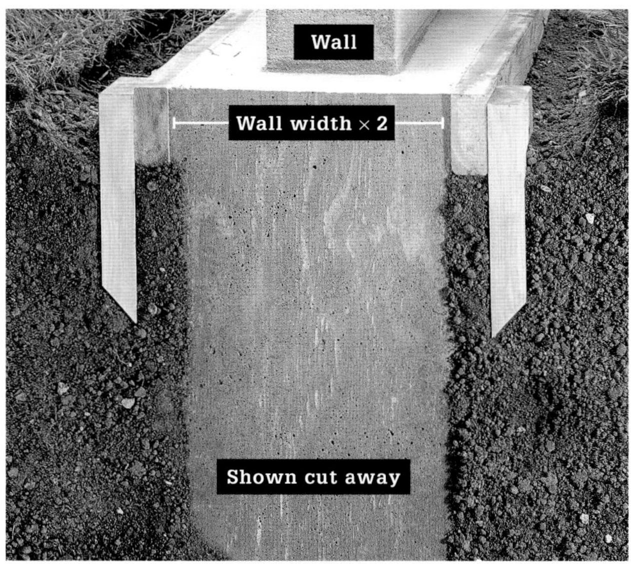

Make footings twice as wide as the wall or structure they will support. They also should extend at least 12" past the ends of the project area.

Add tie-rods if you will be pouring concrete over the footing. After the concrete sets up, press 12" sections of rebar 6" into the concrete. The tie-rods will anchor the footing to the structure it supports.

How to Pour a Footing

Make a rough outline of the footing using a rope or hose. Outline the project area with stakes and mason's string.

Strip away sod 6" outside the project area on all sides and then excavate the trench for the footing to a depth 12" below the frost line.

Build and install a 2 × 4 form frame for the footing, aligning it with the mason's strings. Stake the form in place and adjust to level.

Option: Cut an isolation board and glue it to the existing concrete structures at the point where they meet the new sidewalk. Steps, foundation walls, driveways, and old sidewalk sections are examples of structures you'll need to isolate from the new concrete.

Make two #3 rebar grids to reinforce the footing. For each grid, cut two pieces of #3 rebar 8" shorter than the length of the footing and two pieces 4" shorter than the depth of the footing. Bind the pieces together with 16-gauge wire, forming a rectangle. Set the rebar grids upright in the trench, leaving 4" of space between the grids and the walls of the trench. Coat the inside edge of the form with vegetable oil or commercial release agent.

Mix and pour concrete, so it reaches the tops of the forms. Screed the surface using a 2 × 4. Float the concrete until it is smooth and level.

Cure the concrete for one week before you build on the footing. Remove the forms and backfill around the edges of the footing.

Poured Concrete Walkway

If you've always wanted to try your hand at creating with concrete, an outdoor walkway is a great project to start with. The basic elements and construction steps of a walkway are similar to those of a poured concrete patio or other landscape slab, but the smaller scale of a walkway makes it a much more manageable project for first-timers. Placing the wet concrete goes faster and you can easily reach the center of the surface for finishing from either side of the walkway.

Like a patio slab, a poured concrete walkway also makes a good foundation for mortared surface materials, such as pavers, stone, and tile. If that's your goal, be sure to account for the thickness of the surface material when planning and laying out the walkway height. A coarse broomed or scratched finish on the concrete will help create a strong bond with the mortar bed of the surface material.

The walkway in this project is a four-inch-thick by 26-inch-wide concrete slab with a relatively fine broom finish for providing slip resistance in wet weather. It consists of two straight, 12-foot-long runs connected by a 90-degree elbow. After curing, the walkway can be left bare for a classic, low-maintenance surface, or it can be colored with a permanent acid stain, and can be sealed or left unsealed, as desired. When planning your walkway project, consult your city's building department for recommendations and construction requirements.

Tools & Materials ▸

Drill, bits
Circular saw
Sledgehammer
Mason's string
Line level
Excavation tools
Bow rake
2- or 4-ft. level
Duct tape
Plate compactor or hand tamp
Heavy-duty wire or bolt cutters
Concrete mixing tools
Shovel
Hammer
Magnesium float
Edger
1" groover
Trowel
Push broom
Lumber (2 × 2, 2 × 4)
7/16" hardboard siding or 1/4" lauan for curved forms
Drywall screws (2 1/2, 3 1/2")
4" wood or deck screws
Compactable gravel (3/4" maximum stone size)
6 × 6" welded wire mesh
Metal rebar
Tie wire
2" wire bolsters
Isolation board and construction adhesive
Concrete form release agent
4,000 psi concrete (or as required by local code)
Clear polyethylene sheeting
Eye and ear protection
Sprayer
Work gloves

Poured concrete walkways can be designed with straight lines, curves, or sharp angles. The flat, hardwearing surface is ideal for frequently traveled paths and will stand up to heavy equipment and decades of snow shoveling.

Options for Sloping a Walkway

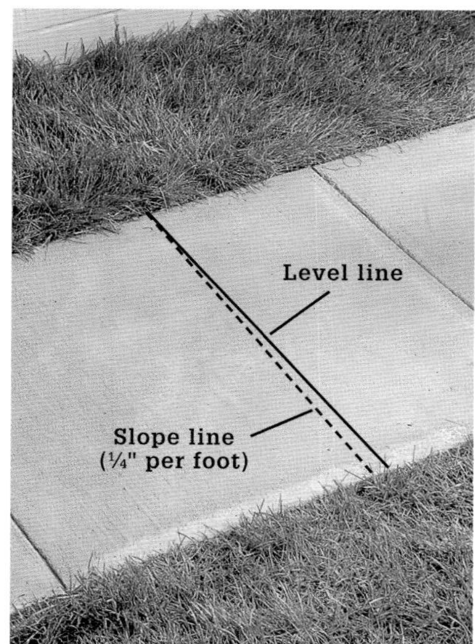

Straight slope: Set the concrete form lower on one side of the walkway so the finished surface is flat and slopes downward at a rate of ¼" per foot. Always slope the surface away from the house foundation or, when not near the house, toward the area best suited to accept water runoff.

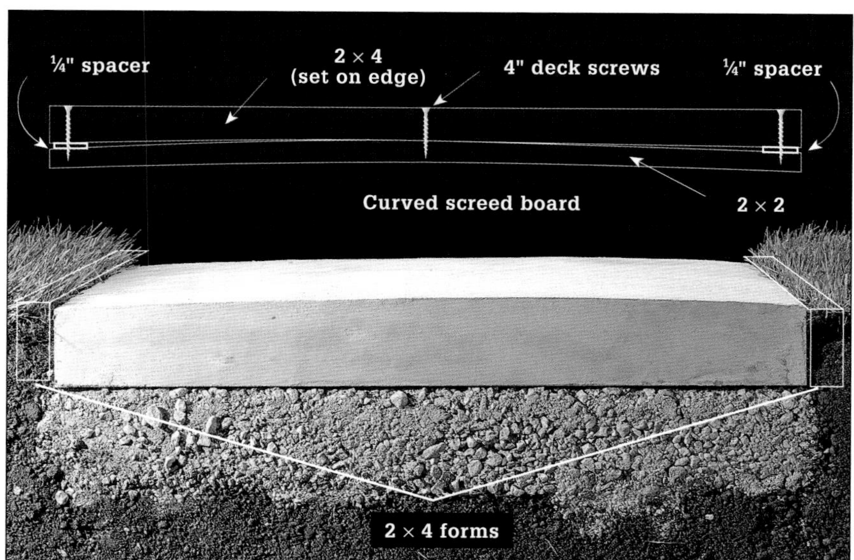

Crowned slope: When a walkway does not run near the house foundation, you have the option of crowning the surface so it slopes down to both sides. To make the crown, construct a curved screed board by cutting a 2 × 2 and 2 × 4 long enough to rest on both sides of the concrete form. Sandwich the boards together with a ¼"-thick spacer at each end, and then fasten the assembly with 4" wood or deck screws driven at the center and the ends. Use the board to screed the concrete (see step 7, page 32).

Options for Reinforcing a Walkway

As an alternative to the wire mesh reinforcement used in the following project, you can reinforce a walkway slab with metal rebar (check with the local building code requirements). For a 3-ft.-wide walkway, lay two sections of #3 rebar spaced evenly inside the concrete form. Bend the rebar as needed to follow curves or angles. Overlap pieces by 12" and tie them together with tie wire. Use wire bolsters to suspend the bar in the middle of the slab's thickness.

How to Build a Concrete Walkway

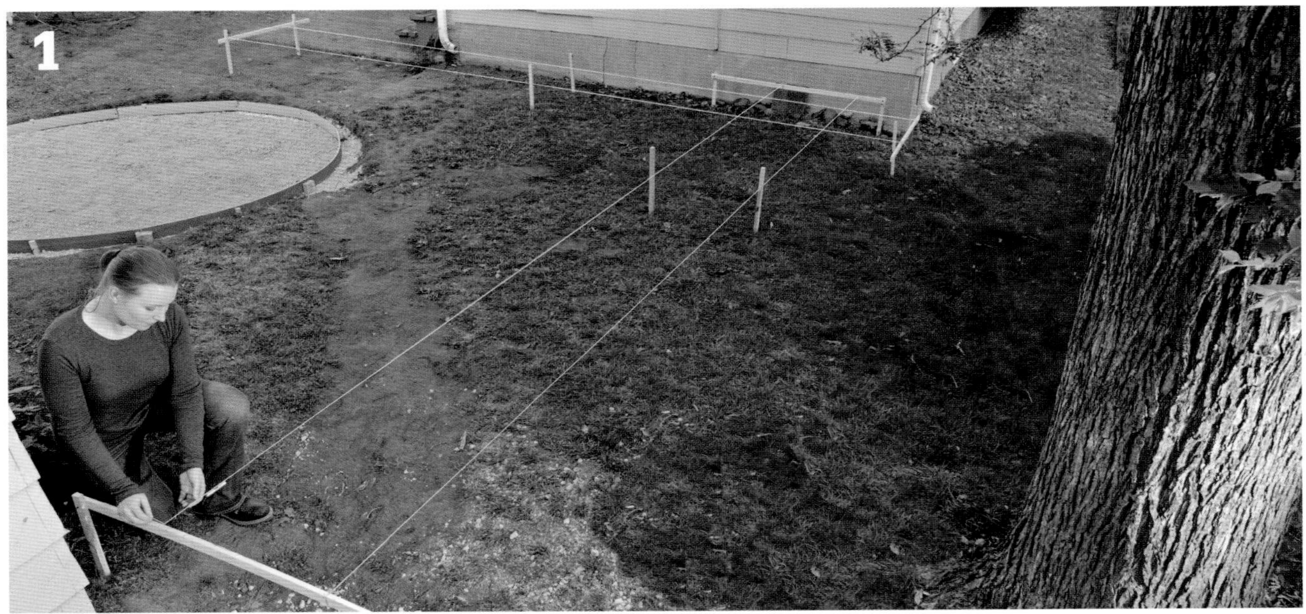

Lay out the precise edges of the finished walkway using stakes (or batterboards) and mason's string (see pages 38 to 39 for additional help with setting up and using layout strings). Where possible, set stakes 12" or so outside of the walkway edges so they're out of the way. Make sure any 90° corners are square using the 3-4-5 measuring technique. Level the strings, and then lower the strings on one side of the layout to create a downward slope of ¼" per foot (if the walkway will be crowned instead of sloped to one side, keep all strings level with one another: see page 39). Begin the excavation by cutting away the sod or other plantings 6" beyond the layout lines on all sides of the site.

Excavate the site for a 6"-thick gravel sub-base, plus any sub-grade (below ground level) portion of the slab, as desired. Measure the depth with a story pole against the high-side layout strings, and then use a slope gauge to grade the slope. Tamp the soil thoroughly with a plate compactor.

Cover the site with a 4" layer of compactable gravel, and then tamp it thoroughly with a plate compactor. Add 4" or more of gravel and screed the surface flat, checking with a slope gauge to set the proper grade. Compact the gravel so the top surface is 4" below the finished walkway height. Reset the layout strings at the precise height of the finished walkway.

Build the concrete form with straight 2 × 4 lumber so the inside faces of the form are aligned with the strings. Fasten the form boards together with 3 ½" screws. Drive 2 × 4 stakes for reinforcement behind butt joints. Align the form with the layout strings, and then drive stakes at each corner and every 2 to 3 ft. in between. Fasten the form to the stakes so the top inside corners of the form boards are just touching the layout strings. The tops of the stakes should be just below the tops of the form.

Add curved strips made from hardboard or lauan to create curved corners, if desired. Secure curved strips by screwing them to wood stakes. Recheck the gravel bed inside the concrete form, making sure it is smooth and properly sloped.

Lay reinforcing wire mesh over the gravel base, keeping the edges 1 to 2" from the insides of the form. Overlap the mesh strips by 6" (one square) and tie them together with tie wire. Prop up the mesh on 2" wire bolsters ("chairs") placed every few feet and tied to the mesh with wire. Install isolation board (see page 37) where the walkway adjoins other slabs or structures. When you're ready for the concrete pour, coat the insides of the form with a release agent or vegetable oil.

Place the concrete, starting at the far end of the walkway. Distribute it around the form (don't throw it) with a shovel. As you fill, stab into the concrete with the shovel, and tap a hammer against the back sides of the form to eliminate air pockets. Continue until the form is evenly filled, slightly above the tops of the form.

(continued)

Immediately screed the surface with a straight 2 × 4: Two people pull the board backward in a side-to-side sawing motion, with the board resting on top of the form. As you work, shovel in extra concrete to fill low spots or remove concrete from high spots, and re-screed. The goal is to create a flat surface that's level with the top of the form.

Option: Cut an isolation board and glue it to the existing concrete structures at the point where they meet the new sidewalk. Steps, foundation walls, driveways, and old sidewalk sections are examples of structures you'll need to isolate from the new concrete.

Float the concrete surface with a magnesium float, working back and forth in broad arching strokes. Tip up the leading edge of the tool slightly to prevent gouging. Stop floating once the surface is relatively smooth and has a wet sheen. Be careful not to over-float, indicated by water pooling on the surface. Allow the bleed water to disappear and the concrete to harden sufficiently (see page 33).

Use an edger to shape the side edges of the walkway along the wood form. Carefully run the edger back and forth along the form to create a smooth, rounded corner, lifting the leading edge of the tool slightly to prevent gouging.

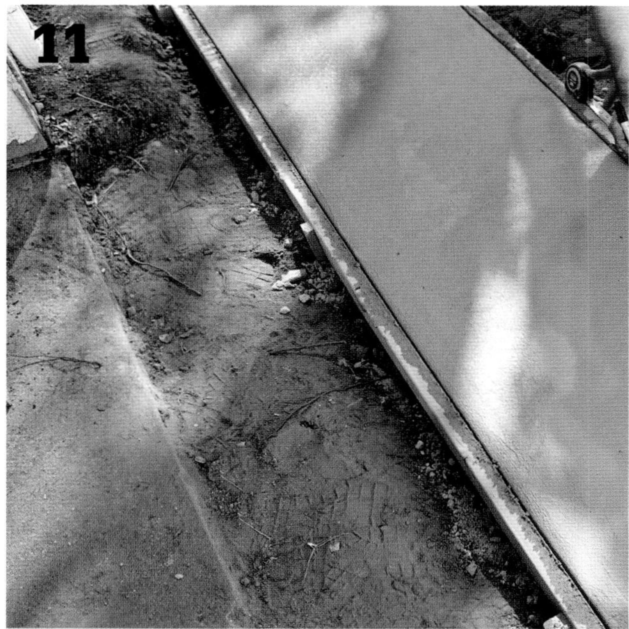

Mark the locations of the control joints onto the top edges of the form boards, spacing the joints 1½ times the width of the walkway.

Cut the control joints with a 1" groover guided by a straight 2 × 4 held (or fastened) across the form at the marked locations. Make several light passes back and forth until the groove reaches full depth, lifting the leading edge of the tool to prevent gouging. Remove the guide board once each joint is complete. Smooth out the tool marks with a trowel or float.

Create a nonslip surface with a broom. Starting at the far side edge of the walkway, steadily drag a broom backward over the surface in a straight line, using a single pulling motion. Repeat in single, parallel passes (with minimal or no overlap), and rinse off the broom bristles after each pass. The stiffer and coarser the broom, the rougher the texture will be.

Cure the concrete by misting the walkway with water, and then covering it with clear polyethylene sheeting. Smooth out any air pockets (which can cause discoloration), and weight down the sheeting along the edges. Mist the surface and reapply the plastic daily for a few days.

Concrete Steps

Designing steps requires some calculations and some trial and error. As long as the design meets safety guidelines, you can adjust elements such as the landing depth and the dimensions of the steps. Sketching your plan on paper will make the job easier. The single-wall plywood forms seen here are sufficient for stairs of this size, but if the scale of your project is larger, add a second layer to each side prevent bowing or blow-out.

Before demolishing your old steps, measure them to see if they meet safety guidelines. If so, you can use them as a reference for your new steps. If not, start from scratch so your new steps do not repeat any design errors.

For steps with more than two risers, you'll need to install a handrail. Ask a building inspector about other requirements.

Tools & Materials ▸

- Tape measure
- Sledge hammer
- Shovel
- Drill
- Reciprocating saw
- Level
- Mason's string
- Hand tamper
- Mallet
- Concrete mixing tools
- Jigsaw
- Clamps
- Ruler or framing square
- Float
- Step edger
- Broom
- 2 × 4 lumber
- Steel rebar grid
- Wire
- Bolsters
- Construction adhesive
- Compactable gravel
- Concrete mix
- Fill material
- Exterior-grade ¾" plywood
- 2" deck screws
- Isolation board
- #3 rebar
- Stakes
- Latex caulk
- Vegetable oil or commercial release agent
- Eye protection and gloves
- J-bolts
- Concrete sealer

New concrete steps give a fresh, clean appearance to your house. And if your old steps are unstable, replacing them with concrete steps that have a non-skid surface will create a safer living environment.

How to Design Steps

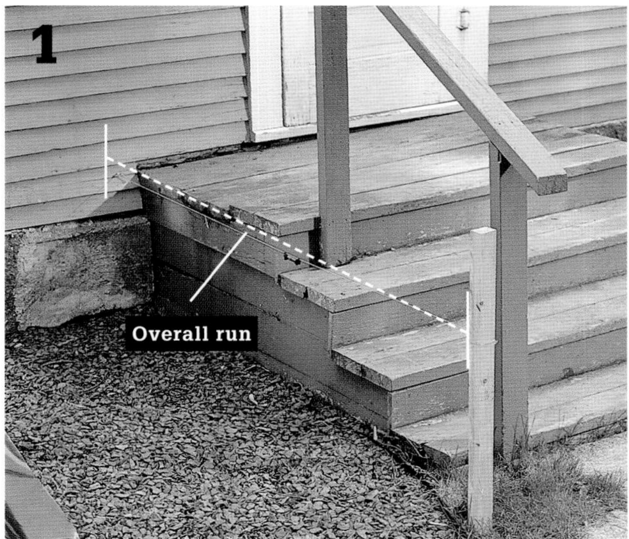

Attach a mason's string to the house foundation, 1" below the bottom of the door threshold. Drive a stake where you want the base of the bottom step to fall. Attach the other end of the string to the stake and use a line level to level it. Measure the length of the string—this distance is the overall depth, or run, of the steps.

Measure down from the string to the bottom of the stake to determine the overall height, or rise, of the steps. Divide the overall rise by the estimated number of steps. The rise of each step should be between 6" and 8". For example, if the overall rise is 21" and you plan to build three steps, the rise of each step would be 7" (21 divided by 3), which falls within the recommended safety range for riser height.

Measure the width of your door and add at least 12"; this number is the minimum depth you should plan for the landing area of the steps. The landing depth plus the depth of each step should fit within the overall run of the steps. If necessary, you can increase the overall run by moving the stake at the planned base of the steps away from the house, or by increasing the depth of the landing.

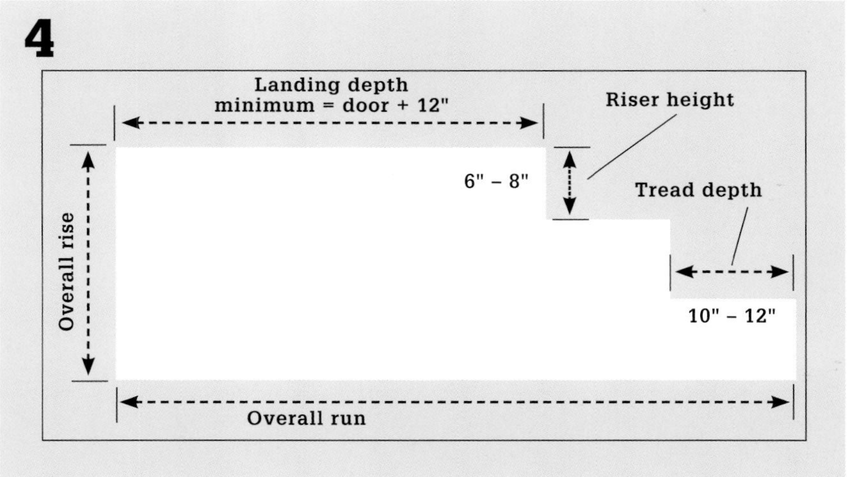

Sketch a detailed plan for the steps, keeping these guidelines in mind: Each step should be 10 to 12" deep, with a riser height between 6 and 8", and the landing should be at least 12" deeper than the swing radius (width) of your door. Adjust the parts of the steps as needed, but stay within the given ranges. Creating a final sketch will take time, but it is worth doing carefully.

How to Build Concrete Steps

Remove existing steps; if the old steps are concrete, break them up and set aside the rubble to use as fill material for the new steps. Wear protective gear, including eye protection and gloves, when demolishing concrete. *Tip: A rental jackhammer can shave hours of hard labor from demolishing concrete steps.*

Dig 12"-wide trenches to the required depth for footings. Locate the trenches perpendicular to the foundation, spaced so the footings will extend 3" beyond the outside edges of the steps. Install rebar grids (page 59) for reinforcement. Affix isolation boards to the foundation wall inside each trench using a few dabs of construction adhesive (12" below the permanent frost line).

Mix the concrete and pour the footings. Level and smooth the concrete with a screed board. You do not need to float the surface afterwards.

When bleed water disappears, insert 12" pieces of rebar 6" into the concrete, spaced at 12" intervals and centered side to side. Leave 1 ft. of clear space at each end.

Let the footings dry for two days, and then excavate the area between them to 4" deep. Pour in a 5"-thick layer of compactable gravel sub-base and tamp until it is level with the footings.

Transfer the measurements for the side forms from your working sketch onto ¾" exterior-grade plywood. Cut out the forms along the cutting lines using a jigsaw. Save time by clamping two pieces of plywood together and cutting both side forms at the same time. Add a ⅛" per foot back-to-front slope to the landing part of the form.

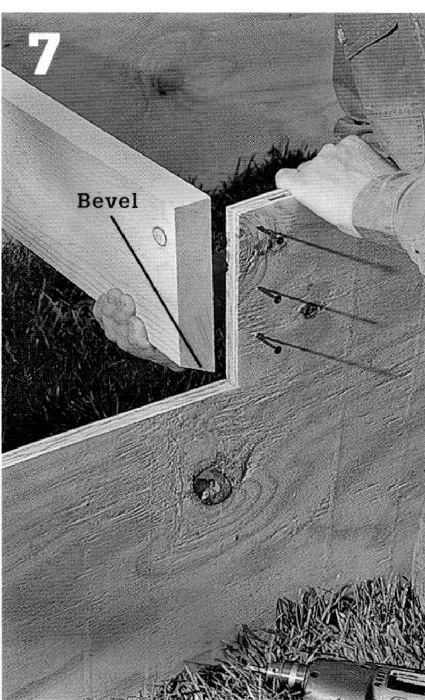

Cut form boards for the risers to fit between the side forms. Bevel the bottom edges of the boards when cutting to create clearance for the float at the back edges of the steps. Attach the riser forms to the side forms with 2" deck screws.

Cut a 2 × 4 to make a center support for the riser forms. Use 2" deck screws to attach 2 × 4 cleats to the riser forms, and then attach the support to the cleats. Check to make sure all corners are square.

Cut an isolation board and glue it to the house foundation at the back of the project area. Set the form onto the footings, flush against the isolation board. Add 2 × 4 bracing arms to the sides of the form, attaching them to cleats on the sides and to stakes driven into the ground.

(continued)

Fill the form with clean fill (broken concrete or rubble). Stack the fill carefully, keeping it 6" away from the sides, back, and top edges of the form. Shovel smaller fragments onto the pile to fill the void areas. This conserves new concrete.

Lay pieces of #3 rebar on top of the fill at 12" intervals, and attach them to bolsters with wire to keep them from moving when the concrete is poured. Keep rebar at least 2" below the top of the forms. Mist the forms and the rubble with water.

Coat the forms with vegetable oil or a release agent so concrete won't stick to the forms. Mix concrete and pour steps one at a time, beginning at the bottom. Settle and smooth the concrete with a screed board. Press a piece of #3 rebar 1" down into the "nose" of each tread for reinforcement.

Float the steps, working the front edge of the float underneath the beveled edge at the bottom of each riser form.

Pour concrete into the forms for the remaining steps and the landing. Press rebar into the nose of each step. Keep an eye on the poured concrete as you work, and stop to float any concrete as soon as the bleed water disappears.

Option: For railings with mounting plates that attach to sunken J-bolts, install the bolts before the concrete sets (page 45). Otherwise, choose railings with surface-mounted hardware (see step 16) that can be attached after the steps are completed.

Once the concrete sets, shape the steps and landing with an edger. Float the surface. Finish by brushing with a stiff-bristled broom for maximum traction.

Remove the forms as soon as the surface is firm to the touch, usually within several hours. Smooth rough edges with a float. Add concrete to fill any holes. If forms are removed later, more patching may be required. Backfill the area around the base of the steps, and seal the concrete. Install a railing.

Entryway Steps

Indoors or outdoors, steps undergo a lot of abuse. Everyday foot traffic, appliance moving, shoveling, and freeze/thaw cycles are just some of the stresses that can contribute to premature step failure. In short, you want steps to be durable. And when it comes to durability, poured concrete is virtually impossible to beat.

Building entryway steps can be a relatively simple landscaping project or it can be a fairly complex home remodeling project involving structural engineers and building permits. The key variables are the total height and whether or not the steps are connected to a permanent structure with a frost footing. If your project is simply two or three concrete steps that solve a slope problem in your yard, the standards are relatively low and you can probably accomplish the project in an afternoon with a couple bags of concrete mix and some gravel. But if your new concrete steps will include three or more risers and will serve as the entryway to your house, then you are looking at a fairly major concrete project.

The concrete steps seen here require enough concrete to make pouring them a fairly big undertaking, but because the structures they are integrated with are a retaining wall and a sandset paver walkway, a full concrete footing that extends beyond the frost line is not required.

Tools & Materials ▸

- Spud bar
- Shovel
- Circular saw
- Power miter saw
- Table saw
- Drill/driver
- Level
- Tamper
- Bow rake
- Wheelbarrow
- Float or trowel
- Compactable gravel
- Crushed rock
- Edging tool
- Lumber (2 × 4, 2 × 8)
- Deck screws
- Concrete release agent
- #3 rebar
- Concrete
- Sheet plastic
- Sprayer
- Concrete stain
- Eye protection and work gloves

Poured concrete steps are hardworking structures that are at home in just about any setting. If they are freestanding (not attached to a house), they normally do not require a frost footing.

Freestanding Stairs Form

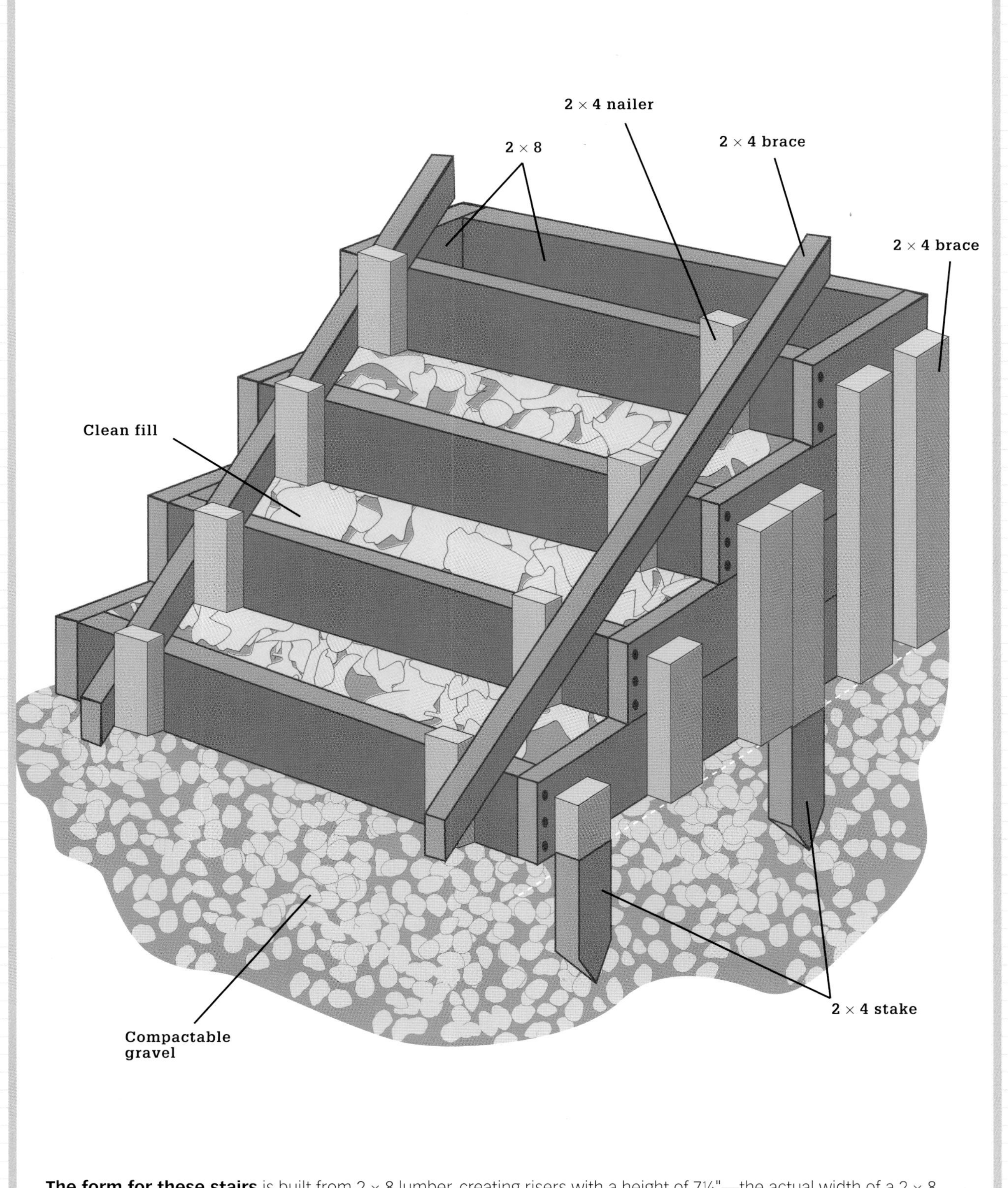

The form for these stairs is built from 2 × 8 lumber, creating risers with a height of 7¼"—the actual width of a 2 × 8.

How to Pour Concrete Landscape Steps

Remove any old steps and prepare the building site. If the old steps are concrete, you'll need to break them up first (see page 58). Wood steps, such as the railroad-tie steps seen here, generally can be pried out with a sturdy spud bar. *Note: Before you do any digging, contact your local utilities to have underground lines identified and flagged.*

Excavate for the new steps. If you are building next to a house, you'll need to dig down past the frost line to pour footings. For larger steps you may find it worth your time to rent a small backhoe or hire an excavation contractor to dig out for the project.

Prepare a sturdy base for the structure by filling the bottom of the excavation area with a 4 to 6" layer of compactable gravel and then compacting it with a hand tamper or gas-powered plate vibrator. The base should extend at least 6" beyond the area where the forms will be located.

Create drainage by adding a 6 to 12" layer of 1 to 2"-dia. crushed rock on top of the compactable gravel. This type of rock is not compactable, but you can settle it somewhat by working it with a bow rake.

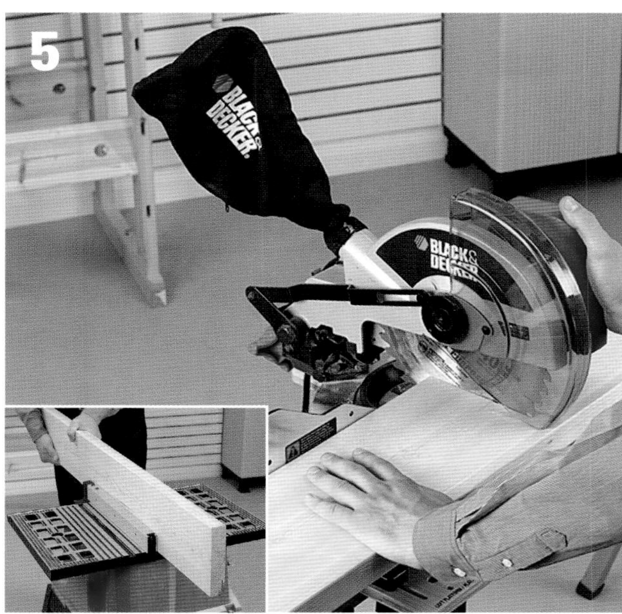

Cut the form boards. Although ¾" plywood is often used for step forms (see pages 59 to 60), 2 × 8 stock is being used here to make the riser forms and the form sides. The 7¼" actual width of a 2 × 8 is perfect for an exterior step riser when coupled with an 11" tread depth, and the beefy thickness of the dimensional lumber makes for a sturdier form that requires less reinforcement and is often easier to strip. To create nosing on the steps, cut a ½ × 1½" rabbet along the top inside edge of each riser form (inset).

Assemble the frames. Here, the form for the steps is being created by building and squaring four three-sided frames that are equal in width. The deepest frame is on the bottom, and the others, which decrease in length by 11" each, are stacked on top of it. When the squared 2 × 8 frames are bound together, they will create the form.

Attach 2 × 4 battens to the sides of the form once the frames are squared, stacked, and pinned together with nails or screws. In addition to binding the frames together, the battens will provide fastening surfaces for the braces that support the form. Attach 2 × 4 riser braces to keep the riser boards from bowing outward.

Set the fully assembled form into the prepared step construction area. You'll need at least one helper for this. Add or remove gravel beneath the bottom of the form until the bottom is at or slightly below grade.

(continued)

Level and brace the form with 2 × 4 braces that are attached to stakes on one end and to the sides of the form battens on the other end. Once the form is level and plumb, reset it so it slopes ⅛" per foot from the back edge of the top step to the bottom on the lowest riser (inset).

Add clean fill inside the form. Chunks of old concrete are perfect for this. The main reason for the clean fill is to conserve on the expense of fresh concrete, but it will also assist in drainage and lower the likelihood of cracking. Once you have filled the form to within (but no closer than) 6" of any surface, dump crushed gravel onto the top of the pile and let it filter down to fill voids.

Place reinforcement inside the form. Cut lengths of #3 rebar about 8" shorter than the step width. Place the rebar strips on top of the clean fill pile so the rocks hold them in place. Position one length a couple of inches back from each crotch where a riser meets a tread. Also place a piece so it reinforces the front of each tread, setting it 1" to 2" back and down from the corner.

Begin filling the form with concrete, starting with the bottom step. Work the concrete with the shovel to help settle out air bubbles and rap the form with a mallet for the same purpose. *Tip: To assure best results, rent a concrete vibrator to settle the material into all the crevices and corners.* Once you have filled each step, immediately strike off the concrete with a float.

Smooth the concrete surfaces once the steps are all filled, working a magnesium float or steel trowel back and forth to level the concrete as you smooth it. Let the concrete set until the bleed water disappears.

Profile the tread noses and sides with a concrete edging tool. Set bolts or hardware for handrails, if required, into the fresh concrete. You can also attach handrail hardware by drilling holes after the concrete has set. For extra traction, broom the treads. Tack sheet plastic over the concrete and let it dry for at least two days.

Remove the forms and finish the steps if desired. Here, the concrete has been colored with concrete stain and a coating of penetrating concrete stain is being applied. Secure the handrail into the stanchions or hardware, if required. Once the coatings are dry, backfill around the steps.

Backfill around the edges of the steps and install or replace pathways and landings at the top and bottom of the landscape steps.

Poured Concrete Slab

A slab foundation commonly used for garages or patios is called a slab-on-grade foundation. This combines a three and a half- to four-inch-thick floor slab with an 8- to 12-inch-thick perimeter footing that provides extra support. The foundation can be poured at one time using a simple wood form.

Because they sit above ground, slab-on-grade foundations are susceptible to frost heave; in cold-weather climates they are suitable only for detached buildings. Specific design requirements also vary by locality, so check with the local building department regarding the depth of the slab, the metal reinforcement required, the type and amount of gravel required for the sub-base, and whether a plastic or other type of moisture barrier is needed under the slab.

The slab shown in this project has a three and one half-inch-thick interior with an eight-inch-wide by eight-inch-deep footing along the perimeter. The top of the slab sits four inches above ground level (grade). There is a four-inch-thick layer of compacted gravel underneath the slab, and the concrete is reinforced internally with a layer of six by six-inch welded wire mesh.

In some areas, you may be required to add more rebar in the perimeter. Check the local code. If this foundation will be used as a garage surface, then set 8-inch-long J-bolts into the slab along the edges, after concrete is poured and finished. These J-bolts are used later to anchor the wall framing to a slab. If this foundation will serve as a patio sub-base, consider using one of the finishing techniques described on pages 238 to 281.

Tools & Materials ▸

- Work gloves and eye protection
- Stakes and boards
- Mason's lines
- Plumb bob
- Shovel
- Tie wire
- Long level
- Tape measure
- Drill
- Wheelbarrow
- Bull float
- Concrete groover tool
- Release agent
- Concrete edging tool
- Wood or magnesium concrete float
- Paint roller
- Compactable gravel
- Lumber (2 × 4, 2 × 8)
- 3" deck screws
- Metal mending plates
- Rewire mesh
- Concrete
- J-bolts
- Concrete cure and seal
- Plate compactor or hand tamper

A concrete slab with an adjoining concrete apron and driveway is the most common garage foundation setup. The same techniques in this project can be applied when pouring a patio or a foundation.

How to Pour a Concrete Slab

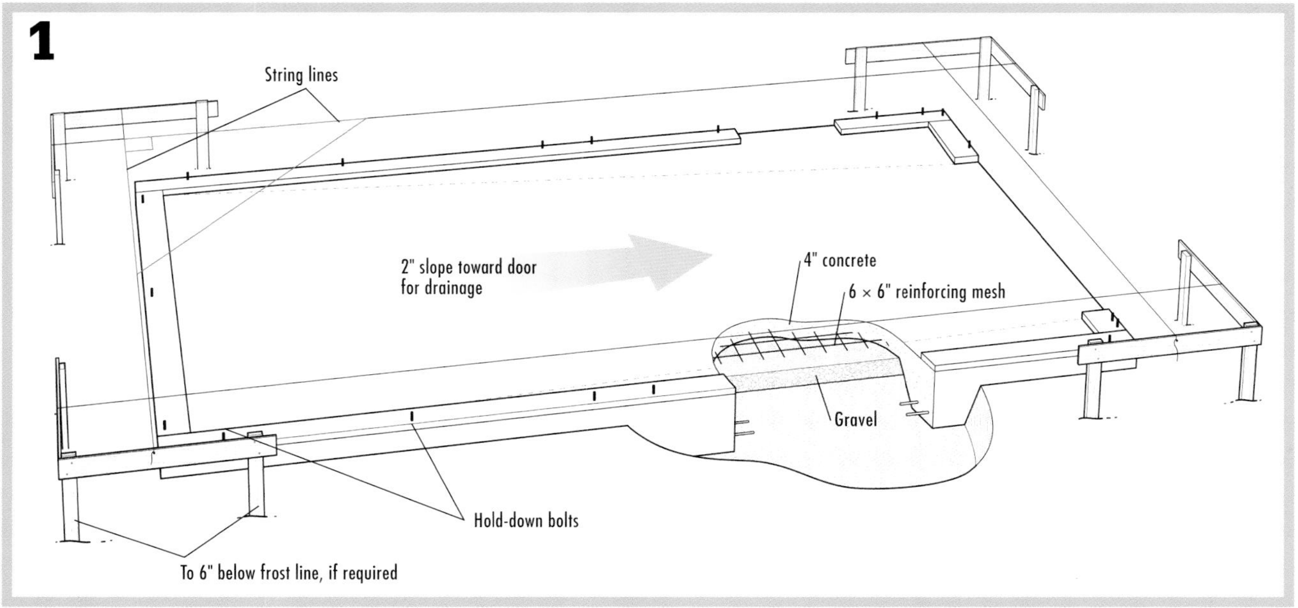

Begin to lay out the excavation with pairs of batterboards installed at each corner of the garage slab site. Position them about 2 ft. outside the perimeter of the slab area so you'll have plenty of room to work. Run level mason's lines between the batterboards to establish the final size of the slab. Drop a plumb bob down from the intersections of the strings and drive a stake at each corner.

Excavate the area about 2 ft. wider and longer than the staked size of the slab. The poured slab should slope slightly to facilitate drainage. Remove 3 to 4" of soil from the excavation area, and dig a deeper trench around the perimeter for the footing. The outside of the footing should line up with the mason's lines. Slope the soil to create a transition between the excavated interior and the footing. Check your local building codes to determine the correct footing size and depth for your climate and soil conditions.

Fill the excavation area with 4" of compactable gravel, letting it spill down into the 12"-deep footings that frame the perimeter. Tamp the gravel level and smooth it with a plate compactor. The gravel surface should maintain the 2" total back-to-front slope. Depending on your soil conditions, some concrete contractors recommend laying 6-mil polyethylene sheeting over the compacted base to form a moisture barrier. *Tip: Install electrical conduit underneath the slab if you will be providing underground electrical service.*

(continued)

Drive woods stakes along the outside of a form (built with 2 × 4 lumber), placing stakes at 4-ft. intervals. Place two stakes at each corner. Set the tops of stakes flush with the top edges of the form (or slightly below the tops). As you drive in stakes, periodically check the form to be sure it is level and measure from corner to corner to ensure that it's square. The form should measure 4" above grade. Attach stakes to the form with deck screws to hold it in place.

Add rewire reinforcement according to the requirements in your area. Here, rows of 6 × 6 wire mesh are set onto spacers (chunks of brick) in the pour area. Overlap the sheets of mesh by 6" and stop the rows about 2" in from the insides of the form. Fasten the mesh together with tie wire. Apply a release agent.

Pour the concrete. Have ready-mix concrete delivered to your job site and place it into the forms with wheelbarrows and shovels (make sure to have plenty of help for this job). Fill a form with concrete, starting at one end. Use a shovel to settle the concrete around the reinforcement and to remove air pockets. Fill the form to the top. *Note: In most municipalities you must have the forms and sub-base inspected before the concrete is poured.*

Strike off the concrete once a section of a form is filled. The best way to do this is to have two helpers strike off (screed) the wet concrete with a long 2 × 6 or 2 × 8 that spans the width of the form. Drag the screed board back and forth along the top of the form in a sawing motion to level and smooth the concrete. Fill any voids ahead of the screed board with shovelfuls of concrete.

Smooth the surface further with a bull float as soon as you're finished screeding, working across the width of the slab. Floating forces aggregate down and draws sand and water to the surface to begin the smoothing process.

Push J-bolts down into the concrete, wiggling them slightly to eliminate air pockets. Twist the bottom hooked ends so they face into the slab. Position the J-bolts 1¾" from the edges of the slab, aligned with your layout marks. Leave 2½" of bolt thread exposed, and make sure the J-bolts are plumb. Smooth the surrounding concrete with a wooden or magnesium concrete float.

Use a magnesium or wood hand-held float to refine the slab's finished surface as soon as the bleed water evaporates. Work the float back and forth, starting from the middle of the slab and moving outward to the edges. Use large scraps of 2"-thick rigid foam insulation as kneeling pads while you work.

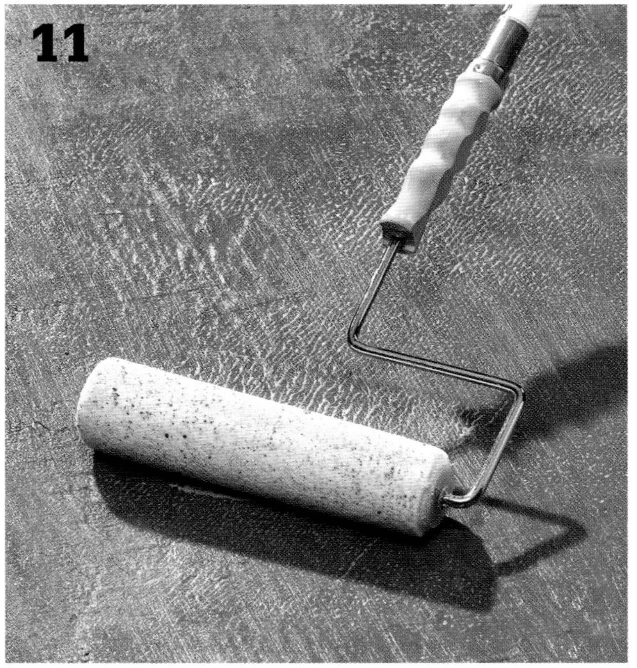

Apply a coat of cure and seal product (See Resources, page 313) to the surface once it dries so you do not have to water the concrete surface during the curing stage. After a couple of days, strip off the forms. Wait at least one more day before you begin building on the slab.

The moldable nature of poured concrete makes it ideal for creating patios with curves and custom shapes in addition to perfect squares and rectangles.

Round Patio

Few outdoor surfaces are as heavy-duty as a properly poured concrete slab. As a patio material, poured concrete is tough to beat. The surface is flat, smooth, easy to clean, and fairly maintenance-free. A concrete slab is also the best foundation for decorative treatments like mortared brick, tile, and stone. And if you like the simplicity and durability of a bare concrete patio, but flat gray doesn't suit your design scheme, you can always apply an acid stain, dry pigment colors, or concrete paint (rated for exterior use) for custom coloring effects that don't compromise the surface's performance.

If you've never worked with poured concrete before, you'll find that most of the time lies in preparing the site and building the forms for containing and shaping the wet concrete. Once the concrete is mixed or delivered to your site, time is of the essence, and the best way to ensure quality results is to be prepared with strong forms, the right tools, and an understanding of each step of the process. And it never hurts to have help: You'll need at least two hardworking assistants for the placing and finishing stages.

This patio project follows the steps for building a small (100 square feet or so) slab that can be poured and finished all at once. The patio featured here is a circular, freestanding structure slightly more than 10 feet in diameter. If you are building a patio of any shape that abuts your house, always isolate it from the house with an isolation board and slope the surface so water drains away from the foundation. A smaller slab is much more manageable for amateurs. Larger slabs often require that you place and tool the wet concrete in workable sections, and these steps must continue simultaneously until the entire slab is filled and leveled before the concrete begins to set. Therefore, it's a good idea to seek guidance and/or assistance from a concrete professional if your plans call for a large patio.

Because they are permanent structures, concrete patios are often governed by local building codes and you might need a permit for your project—especially if the patio abuts a permanent structure. Before you get started, contact your city's building department to learn about permit requirements and general construction specifications in your area, including:

- Zoning restrictions
- Depth of gravel sub-base
- Concrete composition
- Slab thickness and slope
- Internal reinforcement (wire mesh, rebar, etc.)
- Control joints (see page 35)
- Moisture barrier under slab (not a common requirement)

Forming Curves ▸

Creating a smooth curve in a concrete form can be done by using one of several different techniques. The easiest and fastest is to rip-cut strips of bendable sheet stock, such as hardboard, lap siding (nonbeveled), or thin plywood. Use ¼ or ⅜"-thick stock—thinner will flex too much and thicker is difficult to bend. If you need greater rigidity without giving up flexibility, make a curved form by cutting saw kerfs every inch or so about halfway into a piece of 1 × 4. For thicker slabs you may use a 1 × 6.

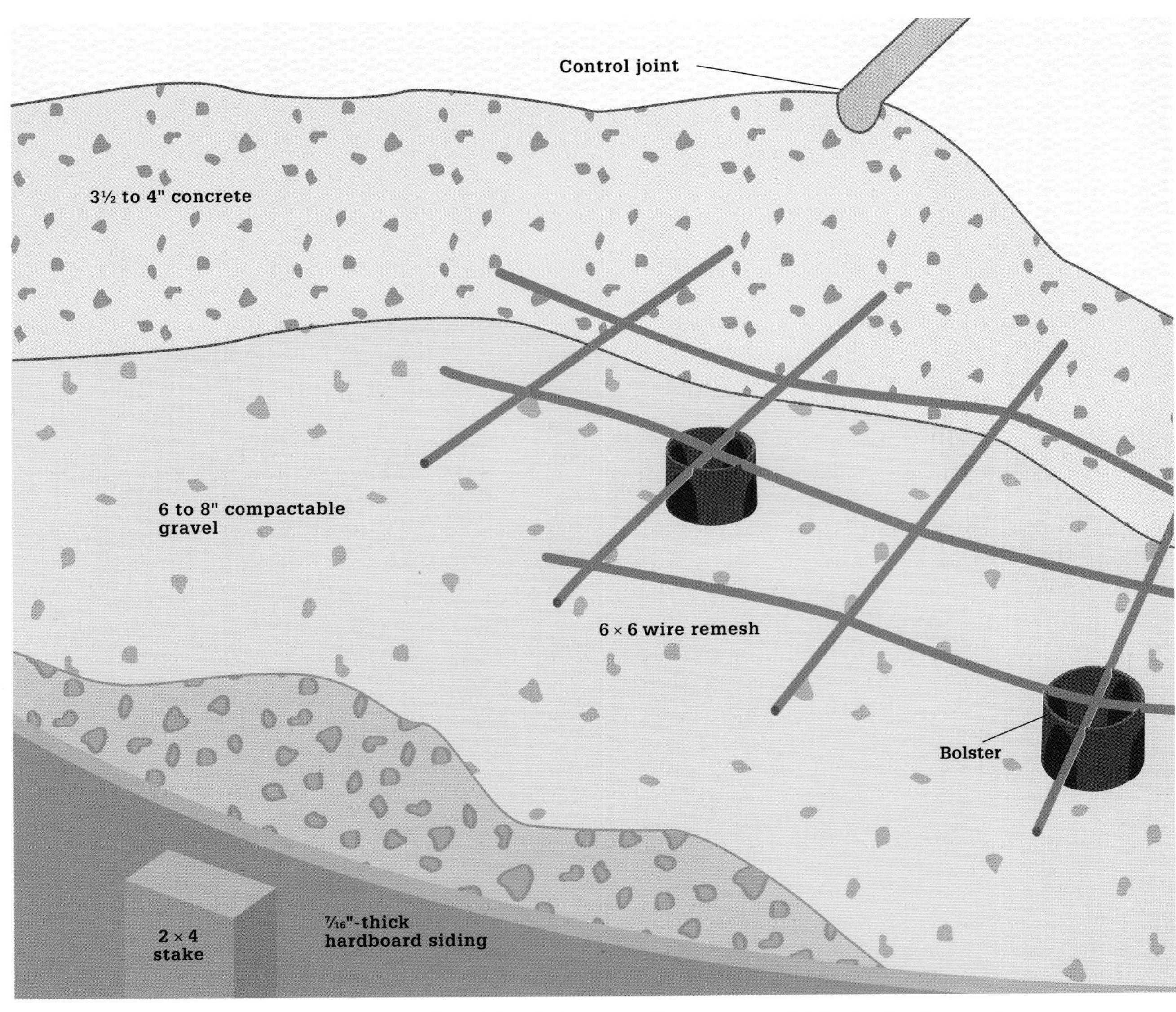

Well-constructed forms and properly prepared foundational elements will ensure your slab is structurally sound.

Tools & Materials ▸

- Drill
- Circular saw
- Hand maul or sledgehammer
- Mason's string
- Stakes
- Marking paint
- Line level
- Excavation tools
- Bow rake
- Level
- Eye protection and gloves
- Lawn edger
- Plumb bob
- Plate compactor or hand tamper
- Chalk line
- Hammer
- 3½" hardboard lap siding
- Heavy-duty wire cutters or bolt cutters
- Concrete mixing tools
- Shovel or masonry hoe
- Wheelbarrow
- Bull float
- Edger
- 1" groover
- Magnesium trowel
- Fine-bristled push broom (optional)
- Lumber (1 × 2, 2 × 4)
- Compactable gravel
- Screws
- 6 × 6" wire mesh
- Tie wire
- 2" wire bolsters
- Release agent
- Isolation board and construction adhesive
- 4,000 psi concrete (or as required by local code)
- Sprayer
- Clear polyethylene sheeting
- Safety protection (per manufacturer recommendations)

How to Build a Round Concrete Patio

Establish layout lines for the site excavation using batterboards, mason's string, and inverted marking paint. Set the lines so they reach at least 12" beyond the work area on all sides. Eventually, the gravel base should extend 12" beyond the slab. Use two pairs of perpendicular batterboards with strings to establish the centerpoint of a round patio (where the strings intersect). To create a rough outline for the patio excavation, drive a stake at the centerpoint and then attach a string to the top of the stake. Tape the other end of the string to a can of inverted marking paint so the distance from the stake to the can equals the radius of the circle, including the gravel base; mark the outline.

Cut the sod on the perimeter of the excavation area to define where to dig. For better access, first remove the batterboards (or at least the strings). A lawn edger works well for cutting the outline into the sod (be sure to wear safety equipment).

Excavate the site for a 6- to 8"-thick compactable gravel sub-base plus any subgrade (below ground level) portion of the slab. If building next to your house, grade the soil so it slopes away from the house at ⅛" per foot. Measure down from leveled cross strings with a story pole to gauge the depth as you work. Compact the soil after grading using a plate compactor or a hand tamper.

(continued)

Patio Next to a House ▸

If your patio will butt up to a house or another permanent structure, you should use the ground level next to the house as your starting point for setting slope and establishing a patio layout. Snap a chalk line onto the house foundation at the precise elevation of the top of the finished slab. This should be 1 to 3" below any patio door threshold. You can use this line for reference during the site prep and concrete pour and finishing.

Fill the excavation area with a 4"-thick layer of compactable gravel. Use an upside-down bow or garden rake to move the rock around. Rake the rock until it is level and follows the grade of the soil base.

Use a plate compactor to tamp the first 4" of graded compactable gravel. Add another 2 to 4" layer of gravel until the top surface is an inch or so above the finished level. Use cross strings and the story pole to make sure the sub-base is uniform and follows the ⅛" per ft. slope. Tamp until the gravel is compacted and at the correct height relative to your lines.

Set level lines for the form height. Replace batterboards and retie the mason's lines so they are level and at the top height of the forms. If you are making a circular patio, as seen here, add intermediate stakes between the batterboards and tie lines to divide the circle into at least eight segments. Drop a plumb bob from the point where the lines intersect and drive a stake at this centerpoint. Use this stake to create a string compass and redraw the patio outline (inset, see step 1).

Drive stakes for anchoring the forms around the perimeter of the patio, just outside the outline. Drive the stakes deep enough that they will be beneath the tops of the forms. Use a hand maul or sledgehammer to drive the stakes. To prevent them from splitting, use a scrap 2 × 4 as a hammer block to absorb the blows. Drive a stake at each point where a string intersects the patio outline.

Install forms. Here, 7⁄16"-thick pieces of hardboard lap siding have been rip-cut into 3½" strips to make bendable forms. Cut each strip long enough to span three stakes as it follows the patio outline. Screw the strip to the middle stake first, making sure the top is flush with the layout string. Bend the form to follow the outline and attach it to the other stakes. Check with a level as you install forms.

Drive stakes behind the forms anywhere where the strips require additional bending or anchoring to follow the round outline. Attach the forms to the stakes. *Note: If you are installing straight 2 × 4 forms, drive screws through the outsides of the stake and into the forms boards to make them easier to remove later.*

(continued)

Lay reinforcing wire mesh over the gravel base, keeping the edges 1 to 2" from the insides of the form. Overlap the mesh strips by 6" and tie them together with tie wire. Prop up the mesh on 2" wire bolsters placed every few feet and tied to the mesh with wire. If required, install isolation board along the house foundation.

Place 4,000 psi concrete in the form, starting at the end farthest from the concrete source. Before pouring, construct access ramps so wheelbarrows can roll over the forms without damaging them, and coat the insides of the form with a release agent or vegetable oil to prevent the forms from sticking. Distribute the concrete with a shovel or masonry hoe. As you fill, hammer against the outsides of the forms to eliminate air pockets.

Screed the surface with a long, straight 2 × 4: Have two people pull the board backward in a side-to-side sawing motion, with the board resting on top of the form. As you work, shovel in extra concrete to fill low spots or remove concrete from high spots and rescreed. The goal is to create a flat surface that's level with the top of the form.

Float the concrete surface with a bull float: Without applying pressure, push and pull the float in straight, parallel passes, overlapping each pass slightly with the next. Slightly tip up the leading edge of the float to prevent gouging the surface. Stop floating once the surface is relatively smooth and has a wet sheen. Be careful not to overfloat, indicated by water pooling on the surface. Allow the bleed water to disappear.

Use an edger to shape all edges of the slab that contact the wood form. Carefully run the edger back and forth along the form to create a smooth, rounded corner. Slightly lift the leading edge of the tool as needed to prevent gouging.

Cut a control joint (if required) using a 1" groover guided by a straight 2 × 4. In most cases, you'll need to erect a temporary bridge to allow access for cutting in the center of the patio. Take great care here. Be sure to cut grooves while concrete is still workable. Make several light passes back and forth until the groove reaches full depth, lifting the leading edge of the tool to prevent gouging.

Flatten ridges and create a smooth surface with a magnesium trowel. This will create a smooth surface that takes a finish well once the concrete has dried. Another finishing option is simply to skip additional floating and let the concrete set up until all the bleed water is gone. Then, brush lightly with a push broom to create a nonslip "broomed" surface.

Cure the concrete by misting the slab with water, then covering it with a single piece of polyethylene sheeting. Smooth out any air pockets (which can cause discoloration), and weight the sheeting along the edges. Mist the slab and reapply the plastic daily for 1 to 2 weeks.

Poured Concrete Wall

Building vertically with poured concrete introduces a whole new dimension to this versatile material. And as much as walls may seem more challenging than slabs or casting projects, the basic building process is just as simple and straightforward. You construct forms using ordinary materials, then fill them with concrete, and finish the surface. While tall concrete walls and load-bearing structures require careful engineering and professional skills, a low partition wall for a patio or garden can be a great do-it-yourself project.

The first rule of concrete wall building is knowing that the entire job relies on the strength of the form. A cubic foot of concrete weighs about 140 pounds, which means that a three-foot-tall wall that is six inches thick weighs 210 pounds for each linear foot. If the wall is 10 feet long, the form must contain over a ton of wet concrete. And the taller the wall, the greater the pressure on the base of the form. If the form has a weak spot and the concrete breaks through (known in the trades as a blowout), there's little chance of saving the project. So be sure to brace, stake, and tie your form carefully.

This project shows you the basic steps for building a three-foot-high partition wall. This type of wall can typically be built on a poured concrete footing or a reinforced slab that's at least four inches thick. When planning your project, consult your local building department for specific requirements, such as wall size, footing specifications, and metal reinforcement in the wall. **Note: This wall design is not suitable for retaining walls, tall walls, or load-bearing walls.**

For help with building a new footing, see pages 48 to 49. The footing should be at least 12 inches wide (2× wall thickness) and at least six inches thick (1× wall thickness), and it must extend below the frost line (or in accordance with the local building code). If your wall will stand on a concrete patio or other slab, the sidebar on page 83 shows you how to install rebar in the slab for anchoring the wall.

Tools & Materials ▸

- Drill and 1/8" bit
- Hacksaw or reciprocating saw
- Pliers
- Level
- Concrete mixing tools
- Shovel
- Concrete trowel
- Lumber (2 × 4, 2 × 2, and 1 × 2)
- 3/4" exterior-grade plywood
- #3 steel reinforcing bar (rebar)
- Hammer or mallet
- Edger
- 16d and 8d nails
- Wood screws or deck screws
- 8-gauge tie wire
- Vegetable oil or commercial release agent
- Concrete mix
- Plastic sheeting
- Exterior-use anchoring cement
- Heavy-duty masonry coating (optional)
- Work gloves and eye protection

In any setting, a poured concrete wall offers clean, sleek lines and a reassuringly solid presence. You can leave the wall exposed to display its natural coloring and texture. For a custom design element, you can add color to the concrete mix or decorate any of the wall's surfaces with stucco, tile, or other masonry finishes.

Wall Form Construction ▸

A wall form is built with two framed sides (much like a standard 2 × 4 stud wall) covered with ¾" plywood. The two sides are joined together at each end by means of a stop board, which also shapes the end of the finished wall. The form is braced and staked in position. Tie wires prevent the sides of the form from spreading under the force of the concrete. Temporary spacers maintain proper spacing between the sides while the form is empty; these are pulled out once the concrete is placed.

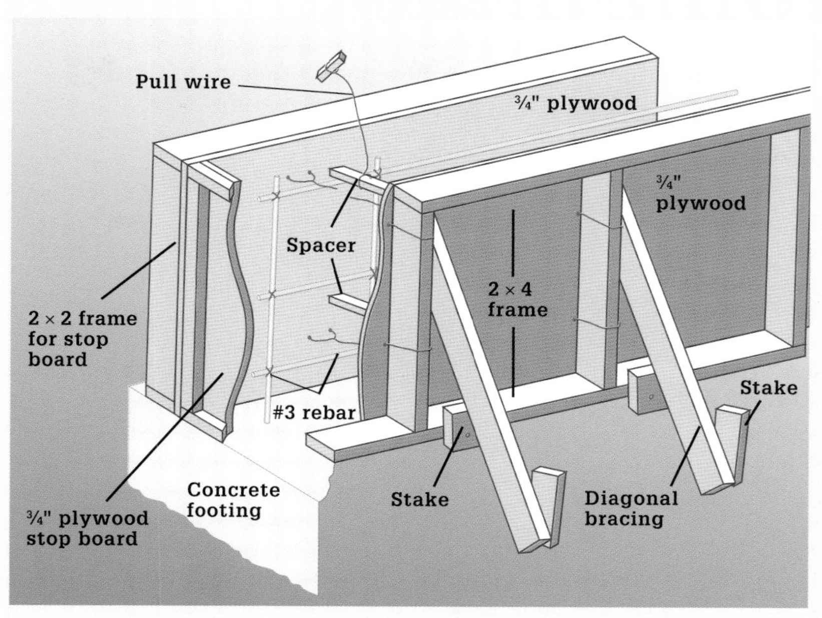

How to Create a Poured Concrete Wall

Build the frames for the form sides from 2 × 4 lumber and 16d nails. Include a stud at each end and every 16" in between. Plan an extra 2¼" of wall length for each stop board. For walls longer than 8 ft., build additional frames.

Cut one piece of ¾" plywood for each side frame. Fasten the plywood to frames with 8d nails driven through the plywood and into the framing. Make sure the top edges of the panels are straight and flush with the frames.

(continued)

Drill holes for the tie wires: At each stud location, drill two pairs of ⅛" holes evenly spaced and keep the holes close to the stud faces. Drill matching holes on the other form side.

Cut #3 rebar at 34", one piece for each rebar anchor in the footing. Cut rebar for three horizontal runs, 4" shorter than the wall length. Tie the short pieces to the footing anchors using 8-gauge tie wire, and then tie the horizontal pieces to the verticals, spacing them 12" apart and keeping their ends 2" from the wall ends. To make a 90° turn, bend the bars on one leg of the wall so they overlap the others by 24".

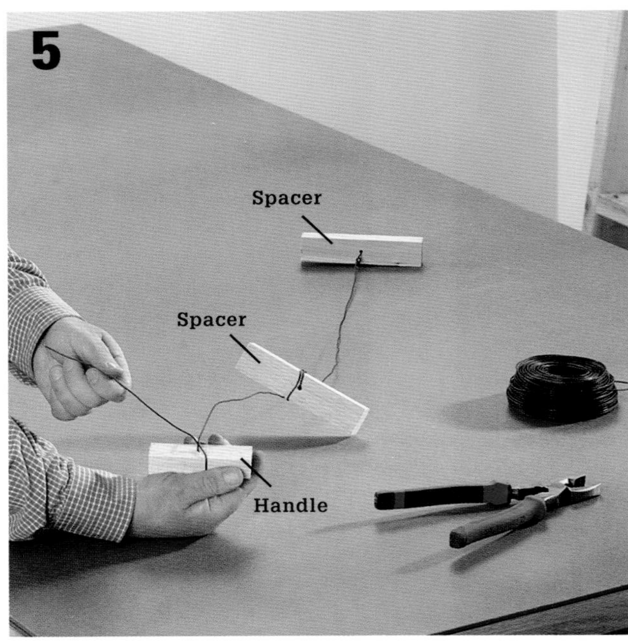

Cut 1 × 2 spacers at 6", one piece for each set of tie wire holes. These temporary spacers will be used to maintain the form width. Tie each pair of spacers to a pull wire, spacing them to match the hole spacing. Then attach a piece of scrap wood to the end of the pull wire to serve as a handle.

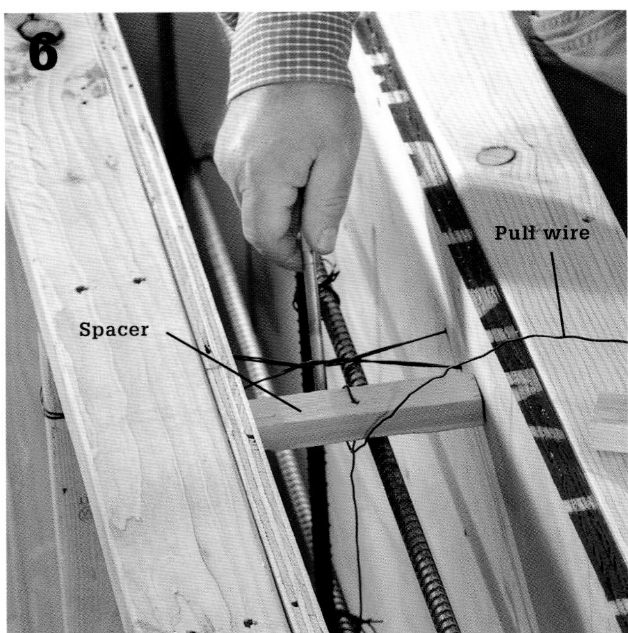

Set the form sides in place. Install the stop boards with 2 × 2 frames for backing; fasten the frames to the form sides with screws. Tie a loop of wire through each set of tie wire holes and position a spacer near each loop. Use a stick to twist the loop strands together, pulling the form sides inward, tight against the spacers.

Building on a Concrete Slab ▸

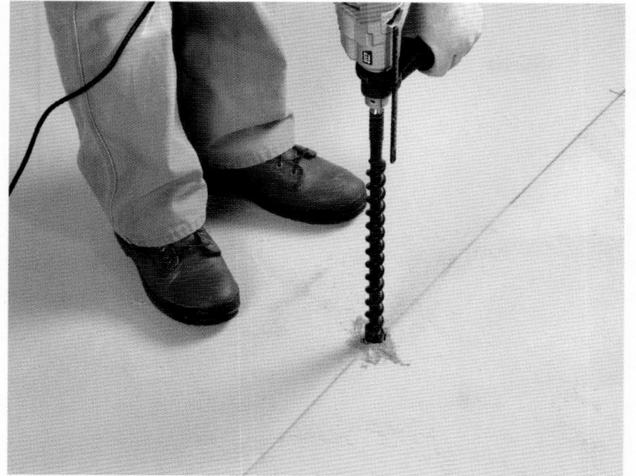

A standard, reinforced 4"-thick concrete slab can be a suitable foundation for a low partition wall like the one shown in this project. The slab must be in good condition, with no significant cracks or changes in level, and you should place the wall several inches away from the slab edge to ensure adequate support. To anchor the new wall to the slab and provide lateral stability, you'll need to install rebar anchors in the slab, following the basic steps shown here. But before going ahead with the project, be sure to have your plans approved by the local building department.

Mark the locations for the rebar anchors along the wall center: Position an anchor 4" from each end of the wall and every 24" in between. At each location, drill a 1½"-diameter hole straight down into the concrete using a hammer drill and 1½" masonry bit (above, left). Make the holes 3" deep. Spray out the holes to remove all dust and debris using an air compressor with a trigger-type nozzle. Cut six pieces of #4 rebar at 16". Mix exterior-use anchoring cement to a pourable consistency. Insert the rods into the holes, and then fill the hole with the cement (above, right). Hold the rods plumb until the cement sets (about 10 minutes). Let the cement cure for 24 hours.

Securing Braces on a Concrete Slab ▸

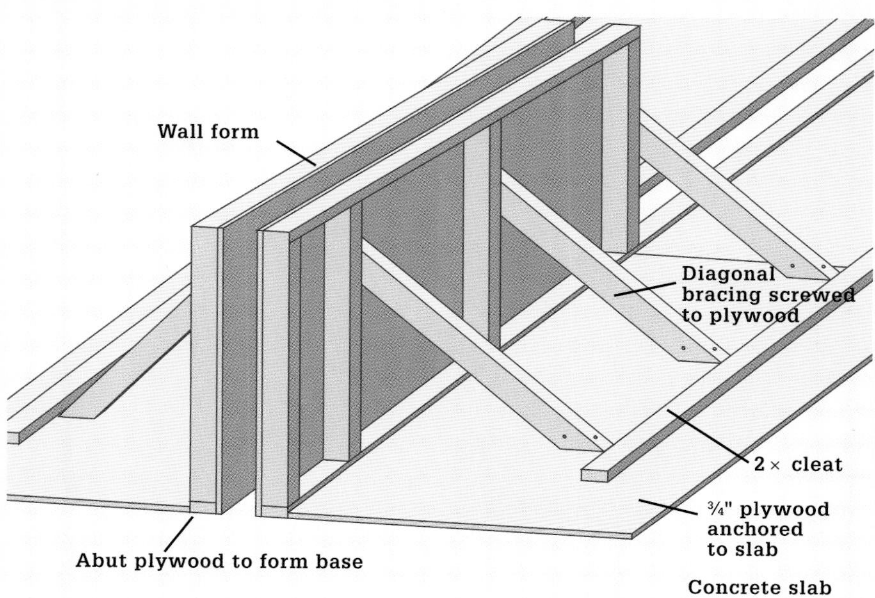

Fasten sheets of ¾" plywood to the slab as an anchoring surface for form braces. Fasten the plywood with a few heavy-duty masonry screws driven into the slab. Butt the sheets against the bottom of the form to provide the same support you would get from stakes. Screw diagonal form bracing directly to the plywood. You can also add a cleat behind the braces for extra support.

(continued)

Make sure the form is centered over the footing. Check that the sides are plumb and the top is level. Secure the form with stakes and braces: Install a diagonal brace at each stud location, and stake along the bottom of the form sides every 12". Fasten all stakes and braces to the form framing with screws. For long walls, join additional side pieces with screws for a tight joint with no gap along the plywood seam. Brace the studs directly behind the joint between sections. Coat the insides of the form with a release agent. If building on a slab (above, right), construct the form and then attach as a unit.

Mix the first batches of concrete in a power mixer, being careful not to add too much water—a soupy mix results in weakened concrete.

Place the concrete in the forms. Start at the ends and work toward the center, filling the form about halfway up (no more than 20" deep). Rap on the forms to settle out air bubbles and then fill to the top. Remove the spacers as you proceed.

Use a shovel to stab into the concrete to work it around the rebar and eliminate air pockets. Continue to rap the sides of the forms with a hammer or mallet to help settle the concrete against the forms.

Screed the top of the wall flat with a 2 × 4, removing spacers as you work. After the bleed water disappears, float or trowel the top surface of the wall for the desired finish. Also round over the edges of the wall with an edger, if desired.

Cover the wall with plastic and let it cure for two or three days. Remove the plastic. Sprinkle with water on hot or dry days to keep concrete from drying too quickly.

Cut the loops of tie wire and remove the forms. Trim the tie wires below the surface of the concrete and then patch the depressions with quick-setting cement or fast-set repair mortar. Trowel the patches flush with the wall surface.

Quick Tip ▸

To achieve a consistent wall color and texture, apply heavy-duty masonry coating with acrylic fortifier using a masonry brush.

Before

Poured Concrete Retaining Wall

Poured concrete has advantages and disadvantages as a building material for structural garden walls, such as this retaining wall. On the plus side: It can conform to just about any size and shape you desire (within specific structural limitations); depending on your source, concrete can be a relatively inexpensive material; poured concrete is very longlasting; with professional engineering, you can build higher with poured concrete than with most other wall materials. But if you live in a region where freeze/thaw cycles exist, you'll need to dig at least a foot past the frost line and provide plenty of good drainage to keep your wall from developing vertical cracks.

A properly engineered retaining wall is designed using fairly complicated dimensional and force ratios. If the wall will be three feet or taller, you should have it engineered by a professional. Shorter retaining walls, sometimes called curb walls, often require less stringent engineering, especially if they are located in a garden setting or are to be used for planting beds or terracing. The wall seen here is built in a fairly cold climate, but the fact that the top is less than 36 inches above ground allows for a drainage base that is above the frost line, with the understanding that some shifting is likely to occur. The project was built in conjunction with poured concrete steps. Because the steps and walls are isolated with an isolation membrane, they are regarded as independent structures and neither is required to have footings that extend below the frost line.

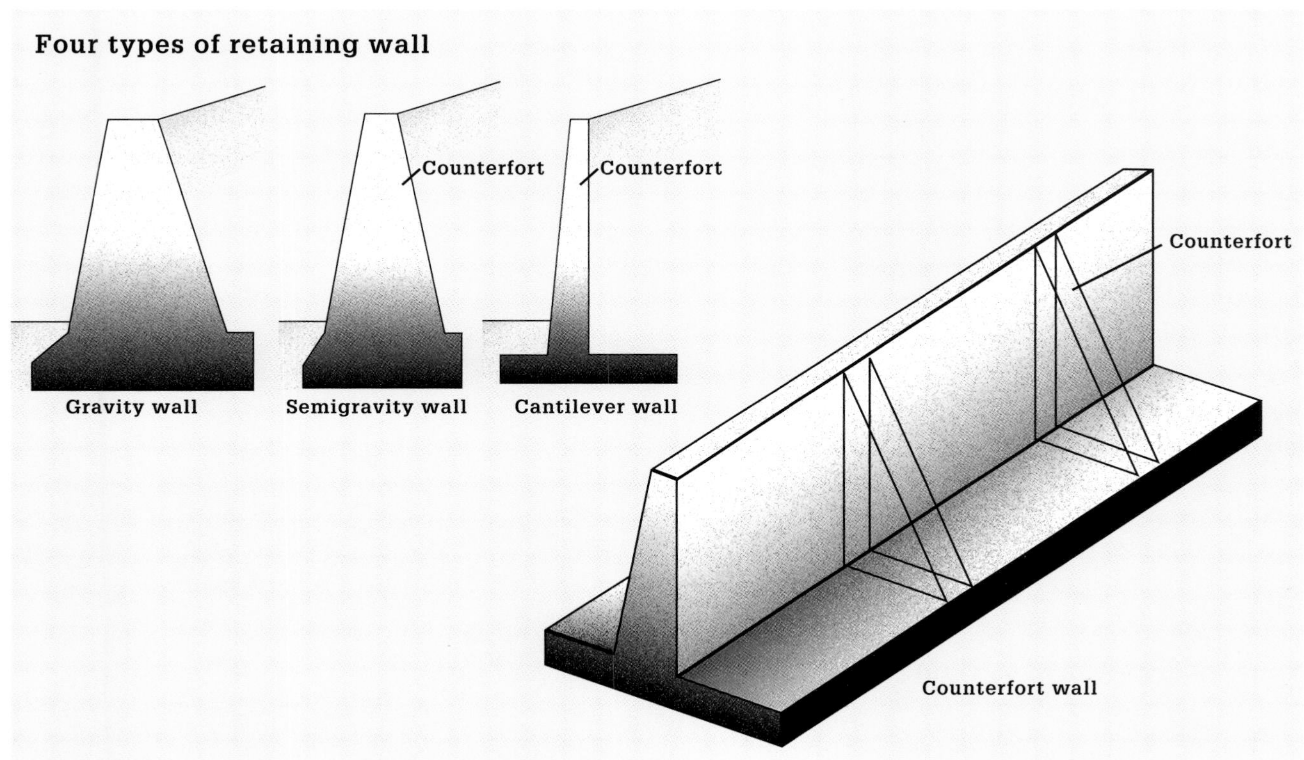

Poured concrete retaining walls employ differing strategies to keep the earth at bay. Some, called *gravity* walls, rely almost exclusively on sheer mass to hold back the groundswell. These are very wide at the bottom and taper upward in both the front and the back. Unless you feel like pouring enough concrete to build a dam, don't plan on a gravity wall that's more than 3 ft. tall. A *semigravity* wall is somewhat sleeker than a gravity wall and employs internal reinforcement to help maintain its shape. A *cantilevered* wall has an integral bottom flange that extends back into the hillside where it is held down by the weight of the dirt that is backilled on top of it. This helps keep the wall in place. A *counterfort* wall is a cantilevered wall that has diagonal reinforcements between the back face of the wall and the flange.

Tools & Materials ▸

- Shovel
- Circular saw
- Power miter saw
- Table saw
- Drill/driver
- Level
- Mason's lines
- Compactable gravel
- Tamper
- Bow rake
- Wheelbarrow
- Float
- Edging tool
- ¾" plywood
- Tie wire
- 2 × 4 lumber
- Deck screws
- Concrete release agent
- #3 rebar
- Concrete
- Sheet plastic
- Concrete vibrator or rubber mallet
- Magnesium trowel
- Edger
- Sprayer
- Concrete stain
- 2" AB plastic pipe
- Landscape fabric
- Drainage gravel
- Eye protection and work gloves
- Plastic sheeting

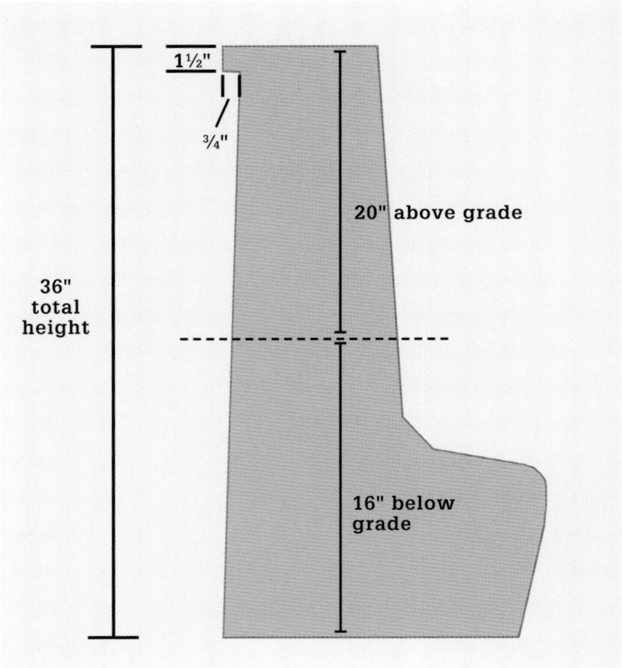

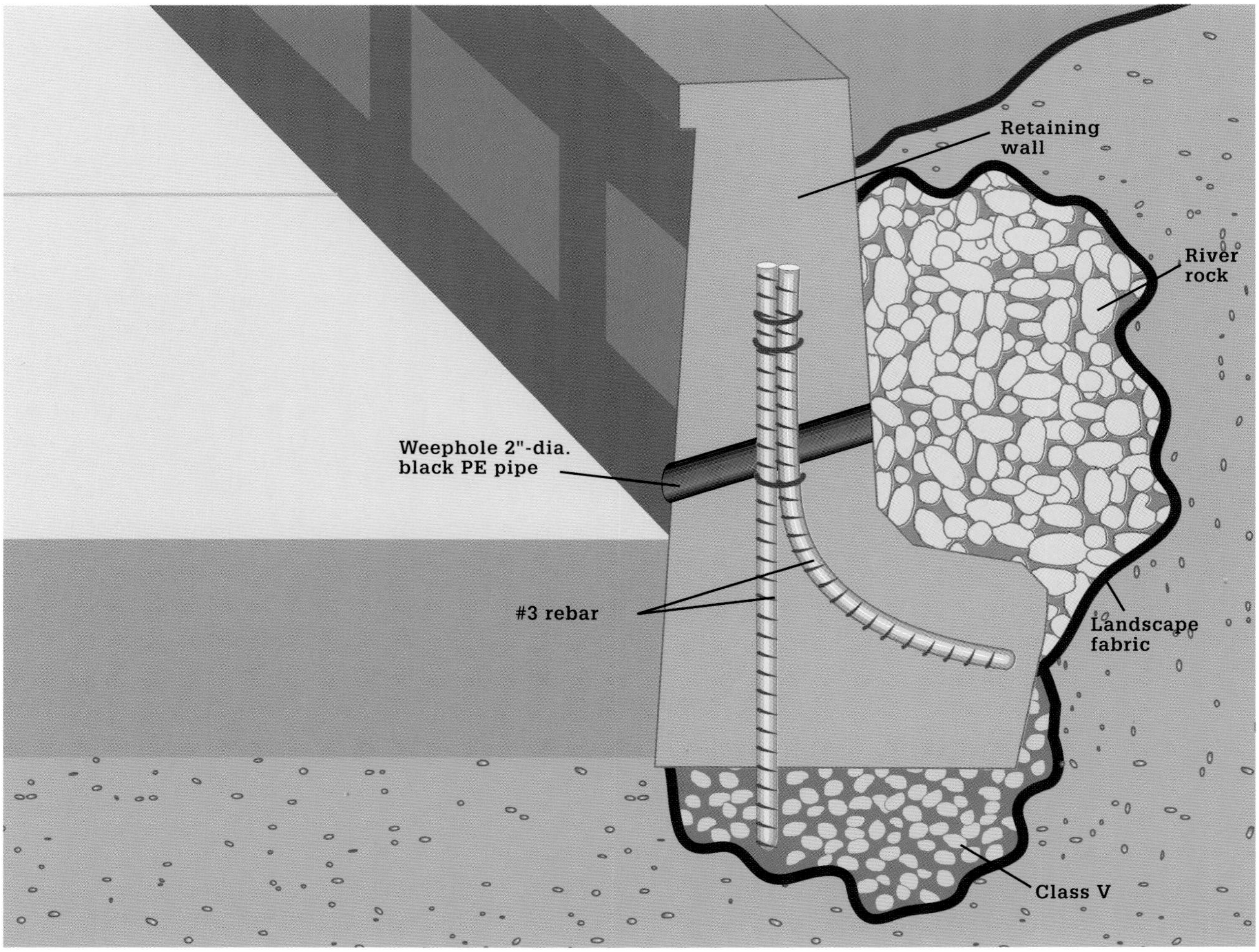

How to Build a Concrete Retaining Wall

Excavate the construction area well beyond the edges of the planned wall. Reserve some soil for backfilling and transport some to lower areas in your yard that need building-up. For larger walls, you can save a lot of work by renting a small backhoe or other earthmoving equipment or hiring an excavator. If your utilities company has flagged any pipes in the area, you must dig around them using hand tools only.

Add a thick layer (at least 4 to 6") of compactable gravel to the bottom of the excavation and tamp it thoroughly with a hand tamper or a rented plate compactor to create a solid foundation for the wall. Add additional base material in sandy or marshy soil.

Cut the form boards, usually from ¾"-thick exterior-grade plywood with one sanded face. You may also use dimensional lumber. Do not use oriented-strand board or particleboard because they have insufficient strength. Do not use any sheet goods that can weaken and delaminate from exposure to wet concrete. Use a circular saw and cutting guide or a table saw to cut panels to width.

Level and stake the forms after doing as much pre-assembly as you can, including attaching the 1 × 4 front forms to create the lip. Run mason's lines at the tops of the forms to use as a reference. Drive 2 × 4 stakes into the ground next to the form braces and attach the braces to the stakes with deck screws. Wherever possible, stake the forms by driving screws through the outer member so they can be removed to facilitate stripping off the forms.

Stake the back forms as well as the front forms. You'll have to get a little creative for this task in many cases, because much depends on the condition of the soil or ground surrounding the wall, as well as access to the forms both during and after the pour. Here, strips of plywood are secured to metal stakes driven into the hill behind the forms. The strips are then screwed to the braces on the back wall forms to hold them in position.

Add rebar reinforcement to tie the integral footing and the wall together. Drive a length of rebar into the bottom of the wall area and then bend another piece and attach it to the rebar stake with wire. Install these reinforcements every 2 to 3 ft. For extra strength, connect them with a horizontal pieces of rebar.

Install weep holes. Choose drain pipe (black AB plastic is a good choice) around 2" in diameter and cut lengths that are equal to the distance between the inside faces of the plywood form boards. Insert the pipes into the bottoms of the form so each end is flush against the inside face of the form. Install a weep hole drain every 6 ft. or so. Drive a long screw through the front panel and the back panel so the screw penetrates the form inside the weep drain, near the top. The ends of the screws will create supports for the drain pipe if the friction fails during the pour (as is likely).

Apply a release agent to the inside surfaces of the forms. You can purchase commercial release agent product or simply brush on a light coat of vegetable oil or mineral oil. Do not use soap (it can weaken the concrete surface) or motor oil (it is a pollutant).

Add Decorative Elements to Forms ▸

One of the beauties of poured concrete is that it is pliable enough to conform to an endless number of decorative forming schemes. For the retaining wall seen here, a series of particleboard panels with beveled edges are attached at regular intervals to the inside faces of the outer form to create a very familiar recessed panel appearance. The panels (beveled edges are cut on a table saw) should be attached with construction adhesive and short screws so there are not gaps that concrete can seep into. Be sure to use a release agent.

(continued)

9

Place the concrete. Most walls call for a 4,000psi mixture (sometimes called 5-bag). You can mix concrete by hand, rent an on-site power mixer, haul your own premixed concrete in a rental trailer, or have the concrete delivered. Begin filling the form at one end of the wall and methodically work your way toward the other end. Have plenty of help so you can start tooling the concrete as soon as possible.

Colored Concrete ▸

The concrete mix seen here was pretinted at the concrete mixing plant. If you do not want a gray concrete structure, using tinted concrete adds color without the need to refresh paint or stain. However, the process is not cheap (about $60 per yard additional), the final color is unpredictable, and you'll have to tint the concrete to match if you need to repair the structure in the future. The pigment that is added can also have unforeseen effects on the concrete mixture, such as accelerating the set-up time.

Hold a panel of sheet stock behind the forms to direct the concrete into the form and prevent it from spilling out.

Settle the concrete in the forms as you work. For best results, rent a concrete vibrator and vibrate thoroughly before screeding. Do not get carried away—overvibrating the concrete can cause the ingredients to separate. A less effective alternative for vibrating (requiring no rental tools) is to work the concrete in with a shovel and settle it by rapping the forms with a rubber mallet.

Strike off, or "screed," the concrete so it is level with the tops of the forms. Use a piece of angle iron on square tubing, or a 2 × 4, as a screed. Move the screed slowly across the forms in a sawing motion. Do not get ahead of the concrete. The material behind the screed should be smooth and level with no dips or voids.

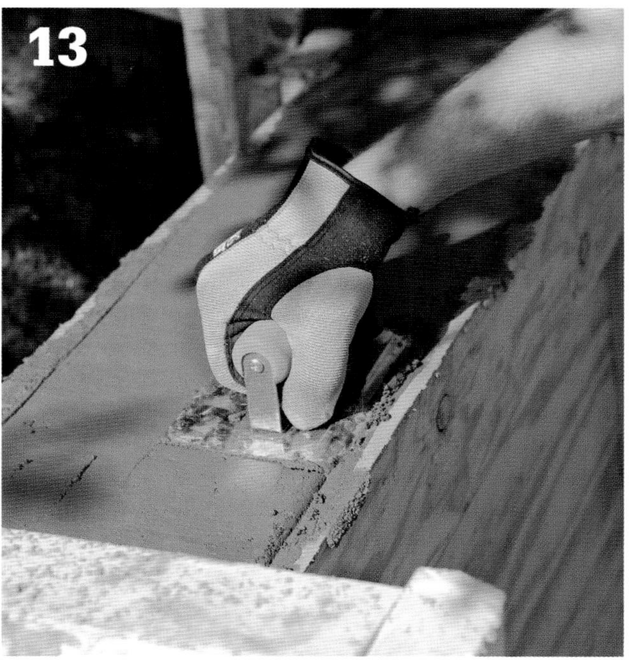

Tool the concrete once the bleed water evaporates, if desired. For a smoother top, float the surface with a magnesium trowel or darby. Run the edger along the top edges on at least the front edge and preferably the back as well.

Cover the concrete surface with plastic sheeting to cure, especially during hot weather. If it is very hot and dry, lift the plastic off and douse the concrete with fresh water twice a day to slow down the drying. Drying too fast can cause cracking and other concrete failures. Wait at least two days before removing the forms.

To backfill, first shovel in an 8 to 12" deep layer of drainage gravel (1 to 2" dia.), and then place a layer of landscape fabric over the gravel to keep dirt out. Shovel dirt over the gravel and tamp it lightly until the desired grade is achieved.

Cast Concrete

Casting decorative objects for the home and yard is an entertaining and creative exercise in handling concrete. Locating and making forms is a challenge itself, and the specific nature of the objects you eventually cast is often dictated by the potential you see in everyday objects that you encounter.

The best forms for casting are rigid or semirigid with a slick surface and the ability to contain water. Plastic and rubber objects are ideal, but you can really use just about any material, especially if you use it for a single casting and are not concerned about breaking it when you release the cast object. Some examples of useful "found" forms include five-gallon plastic buckets (insert a smaller bucket or a tube to cast a large concrete pot); trash can lids (pavers); nesting plastic bowls (pots and planters); or any sphere shapes, such as a basketball, that can be split in two (decorative orbs, bowls).

Making your owns forms is another fun exercise in creativity. Melamine-coated particleboard is a great material for this job because it holds its shape and the concrete will not stick to the surface. When combined with other materials such as the metal flashing used to form the patio tabletop in this chapter, your casting options are practically unlimited.

For more complex and sophisticated castings, you can buy reusable forms in a very wide array of shapes and sizes. Garden benches, birdbaths, landscape edging, pavers, and statuary are just some of the objects you can cast with a couple bags of concrete and a purchased form. The best source for concrete casting forms is the Internet (see Resources section).

One of the best reasons to cast your own decorative and functional objects from concrete is that you can customize the finished look by coloring the concrete or using creative surface treatments such as the footsteps in the pavers seen above.

Casting concrete is a DIY-friendly way to gain experience with handling concrete. No matter your masonry skills, there is a casting project that will challenge and reward you. Projects can be made using purchased molds or everyday objects.

Introduction to Casting

Many objects can be used as forms for casting concrete. Semirigid items work best because they require no special preparation, often include embossed shapes or designs, and are usually easy to strip from the cast object.

Once you try casting, you'll begin to think of every inanimate object as a potential "form" for a concrete structure. Start with simple, easy-to-cast forms such as plastic bowls or buckets. Choose forms that have interesting textures. Consider how you will use the cast concrete when the project is complete: Will it be functional or decorative? You may decide to create a form to suit your needs

Constructed forms can cast items of practically any size or shape. For flat items, such as tabletops and stepping stones, melamine-coated particleboard is a good choice because the melamine side does not require a release agent. Wood should be treated with a release agent like nonstick cooking spray.

You'll need to reinforce large concrete objects with metal rods and mesh. Be sure to keep metal away from the edges. Mix in synthetic reinforcing fibers if not included in your mix. You can purchased bagged concrete with fiber reinforcement blended in at the plant. These products are ideal for large castings that will undergo stress, such as stepping stones or tabletops. For small items, use sand mix. In both cases, add acrylic fortifier to make the mixture more slippery without decreasing strength. Add a coloring agent (liquid or dry) to the mixture to enhance the finished piece.

Sample Casting Projects

A deck bowl is cast using two nesting mixing bowls. This technique can be used with plastic bowls and buckets and planters of all sizes. Larger containers should be split in half and taped back together so you can extract the cast object more easily. We used sand mix with acrylic fortifier and black concrete pigment for this deck bowl.

Sand casting is a great way to use up the leftovers from a larger poured concrete project. To make this birdbath, you simply pile up some coarse wet sand and pour the leftover concrete onto the pile. Birds love the rough texture of the concrete surface.

Garden Column

Prefabricated concrete casting forms give you the ability to make objects for your yard and garden that rival the best (and very expensive) artwork pieces sold at garden centers. Garden benches and birdbaths are among the most popular, but you can locate an array of forms for just about any objects you can imagine.

Because most of the objects cast with readymade forms feature grooves, flutes, or complex patterns, you'll have the best luck if you use a relatively wet mixture of concrete with small or sand-only aggregate. Adding latex bonding agent or acrylic fortifier also makes the concrete more slippery so it can conform to odd shapes more readily, but these agents do not reduce concrete strength, as adding more water does.

If your cast project will be placed outdoors, apply a penetrating concrete sealer about a week after the casting.

Tools & Materials ▸

- Shovel
- Mortar box
- Concrete forms
- Bagged concrete mix (fiber reinforced)
- Acrylic fortifier
- Nonstick cooking spray
- Duct tape
- Exterior landscape adhesive

Prefabricated casting forms typically are made from rugged PVC so they may be reused many times. You can mix and match the forms to create different objects. The forms above include a column form with grapevine or fluted insert, two different pedestal shapes, and an optional birdbath top.

This classical concrete column is cast using a simple plastic form purchased from an Internet supplier (see Resources). It can be used to support many garden items, including a display pedestal, a birdbath, or a sundial.

How to Cast a Garden Column

Choose a column form insert (optional) and slide it into the column form as a liner so the edges meet neatly. Tape the column together at the seam. Coat the insides of all form parts with a very light mist of nonstick cooking spray as a release agent.

Choose a sturdy, level work surface. Set the column form upright on a small piece of scrap plywood. Tape down the form with duct tape, keeping the tape clear of the form top. Mix a batch of fiber-reinforced concrete with an acrylic fortifier and shovel it into the forms. Rap the forms with a stick to settle the concrete and strike off the excess with a screed. Run additional tape "hold-downs" over the top of the column form to secure it to the plywood scrap tightly enough that the concrete will not run out from the bottom.

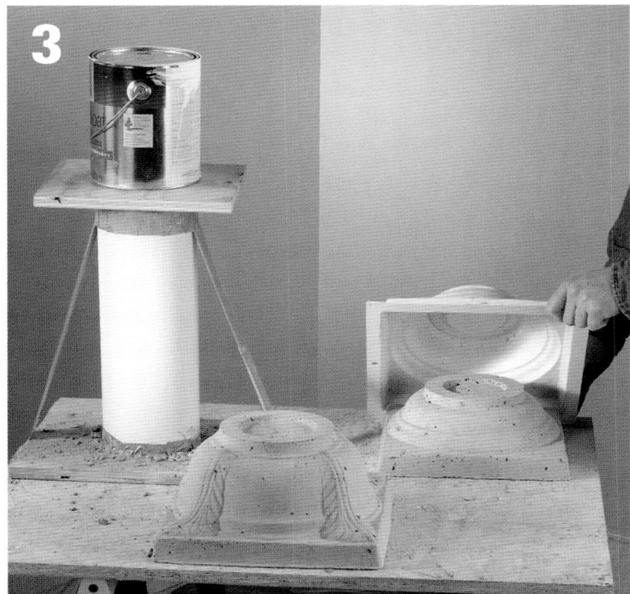

Set another scrap of plywood onto the top of the column form and weight it down. Let the parts dry for two days and then release them from the forms. Wash and rinse the parts to remove dusty residue.

Apply exterior landscape adhesive to the top of the base pedestal and set the column end into the adhesive so the column is centered. Bond the top pedestal in the same manner. Apply penetrating sealant. If it is not nearby, transport the column and pedestals to the location before bonding the parts.

Patio Tabletop

Casting concrete is a good way to produce some types of replacement parts, such as a new top for this old iron patio table base. To make the form for this project, a strip of galvanized roll flashing is inserted inside a ring of finish nails to create a circular shape. Larger tabletops should have rebar or rewire reinforcement, but this 24-inch-diameter top is small enough that fiber reinforcement strands are sufficient.

This round patio tabletop was cast with fiber-reinforced concrete tinted yellow. It is a much simpler version of the kitchen island countertop cast on pages 100 to 107. By casting your own top, you can custom fit any table base you may have.

Tools & Materials ▸

- Aviator snips
- Hammer
- Caulk gun
- Magnesium float
- Pencil
- Drill/driver
- Rubber mallet
- Hammer
- Reinforced concrete mix
- Acrylic fortifier
- Concrete colorant
- Galvanized metal roll flashing
- 6d nails
- Caulk
- Duct tape
- Melamine-coated particleboard
- Tabletop hardware
- String
- Work gloves

How to Cast a Round Tabletop

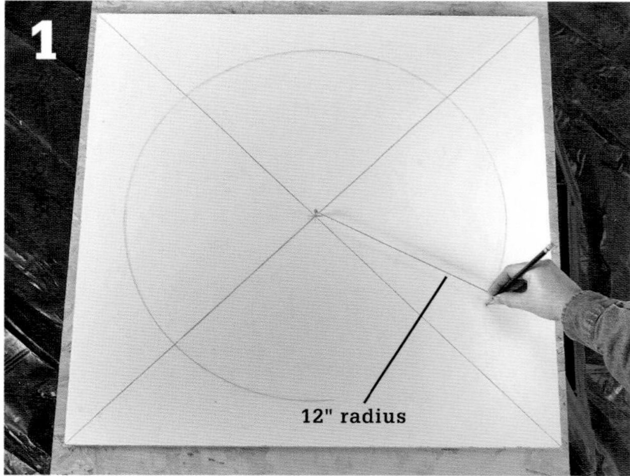

Cut a piece of ¾" melamine (or just about any other sheet stock) to 30 × 30" and drive a small nail in the exact center. Tie string to the nail and tie a pencil to the other end, exactly 12" away form the nail. Pull the string tight and use this "compass" to draw a 12" radius (24" diameter) circle.

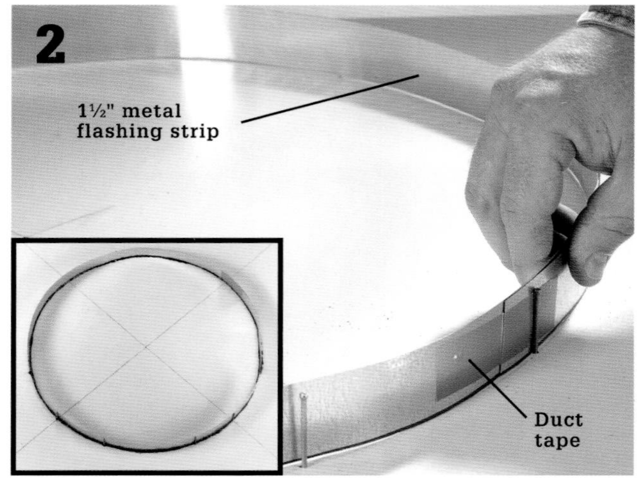

Drive 6d finish nails on the circle line at 6" intervals. Keep the nails perpendicular. Cut a 1½" wide by 80"-long strip of galvanized (not aluminum) flashing using aviator snips. Fit the flashing inside the circle with the cut edge down (factory edge up). Let the flashing spring out inside the circle and adjust so the circle is even. Tape the ends of the flashing with duct tape on the outside of the form.

Caulk liberally around the outside edge of the flashing where it meets the base. After the caulk dries, fill the form with a relatively stiff mix of fiber-reinforced concrete. Add acrylic fortifier and pigment (if desired) to the water before mixing the concrete. Pound the form base gently with a rubber mallet to settle the concrete. Also, work a stick around the edges of the concrete form to help settle the concrete and make sure there are no voids.

Strike off the the top of the concrete with a wood screed and fill any voids or dips in the surface with fresh concrete. Let the concrete set up until the bleed water that rises to the surface evaporates. Then, float the surface with a wood or magnesium float to create a smooth, hard surface. Do not overwork the float, however, as this will draw aggregate to the surface and weaken the pour.

Let the concrete cure for at least two days, then remove the flashing, and release the tabletops. Grind and polish if you wish (see page 107). Then, attach the top to the table base. The table base shown here has screwholes in the bearing ring, so we marked the screw locations on the underside of the tabletop and then drilled holes for screw anchors. Seal the surface.

Kitchen Countertop

Cast concrete countertops have many unique characteristics. They are durable, heat resistant, and relatively inexpensive (if you make them yourself). But most of all, they are highly attractive and a great fit in contemporary kitchens or bathrooms.

A concrete countertop may be cast in place or formed offsite and installed like a natural stone countertop. Casting offsite makes more sense for most homeowners. In addition to keeping the mess and dust out of your living spaces, working in a garage or even outdoors lets you cast the countertops with the finished surface face down in the form. This way, if you do a careful job building the form, you can keep the grinding and polishing to a bare minimum. In some cases, you may even be able to simply remove the countertop from the form, flip it over, and install it essentially as is.

Tools & Materials ▸

Tape measure
Pencil
Table saw or circular saw
Jigsaw
Drill and right-angle drill guide
Level
Carpenter's square
Reciprocating saw with metal-cutting blade
Aviation snips
2" coarse wallboard screws
Deck screws (3, 3½)
Wire mesh
Pliers
Concrete mixer
5-gal. buckets
Shovel
Wheelbarrow
Wooden float
Variable speed angle grinder with grinding pads
Belt sander
Automotive buffer
Insulation board
Plastic sheeting
Rubber mallet
Welded wire mesh for reinforcement
Black or colored silicone caulk
Grinding and polishing pads
Melamine-coated particleboard for constructing the form
Concrete sealer
Coloring agent (liquid or powder)
Compass
#3 rebar
Tie wire
Panel or silicone adhesive
Bagged concrete countertop mix or high/early mix rated for 5,000 psi
Paste wax
Work gloves and eye protection

If installing sink:
Knockout for faucet
Buffing bonnet for polisher
Faucet set
Sink
Polyurethane varnish

Building a custom concrete countertop like this is an easier project than you might think. All of the building materials and techniques are covered in this book.

Planning a Concrete Countertop ▸

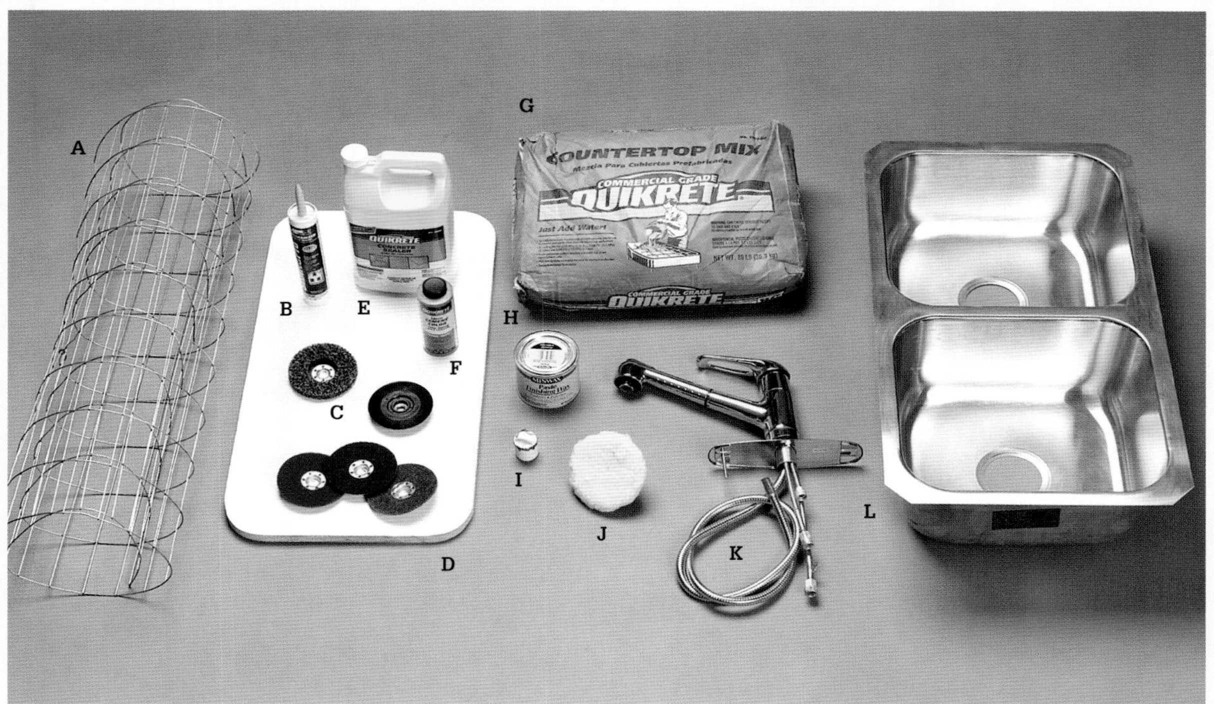

The basic supplies needed to build your countertop form and cast the countertop include: (A) welded wire mesh for reinforcement; (B) black or colored silicone caulk; (C) grinding and polishing pads; (D) melamine-coated particleboard for constructing the form; (E) concrete sealer; (F) coloring agent (liquid or powder); (G) bagged concrete countertop mix or high/early mix rated for 5,000 psi; (H) paste wax; (I) knockout for faucet, if installing sink; (J) buffing bonnet for polisher; (K) faucet set; and (L) sink.

Custom Features: Concrete countertops are normally cast as flat slabs, but if you are willing to put a little more time and effort into it, there are many additional features you can create during the pour. A typical 3"-tall backsplash is challenging, but if you have room behind the faucet you can create a $\frac{3}{4}$"-tall backsplash shelf in the backsplash area. Or, if you search around for some additional information, you can learn how to cast a drainboard directly into the countertop surface. And there is practically no end to the decorative touches you can apply using pigments and inserts.

Estimating Concrete for Countertops: After you design your project and determine the actual dimensions, you'll need to estimate the amount of concrete you'll need. Concrete is measured by volume in cubic feet; multiply the length by the width and then by the thickness of the finished countertop for volume in cubic inches, then divide the sum by 1,728 for cubic feet. For example, a countertop that will be 48" long × 24" deep × $3\frac{1}{2}$" thick will require $2\frac{1}{3}$ cu. ft. of mixed concrete (48 × 24 × 3.5 / 1,728 = $2\frac{1}{3}$) or four 80-lb. bags of countertop mix.

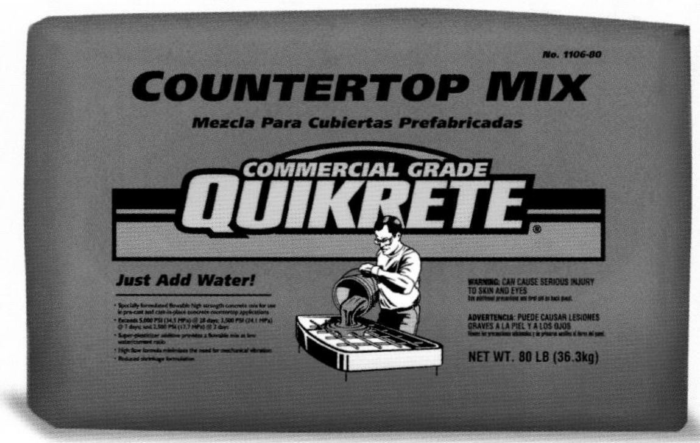

Countertop mix is specially formulated concrete countertop mix designed for use in either precast or cast-in-place projects. Countertop mix contains additives that improve the workability, strength, and finish of the mix.

How to Cast a Concrete Countertop

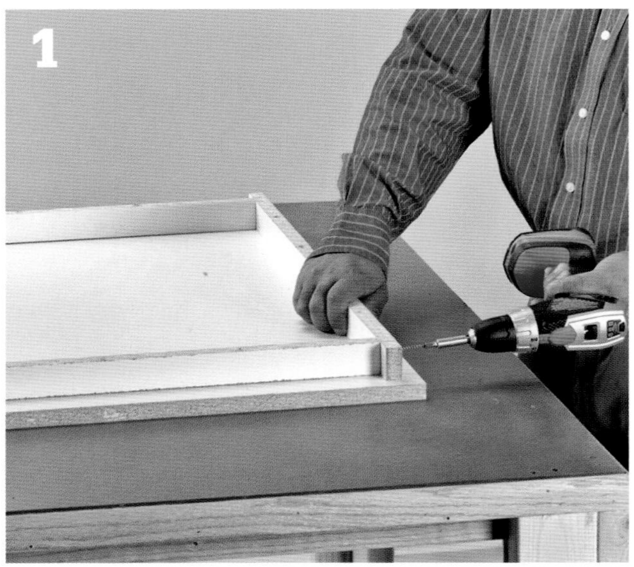

Make the form parts. First, cut 1½"-wide strips of ¾" melamine-coated particleboard for the form sides. Cut the strips to length (26 and 81½" as shown here) and drill two countersunk pilot holes ⅜" in from the ends of the front and back form sides. Assemble the strips into a frame by driving a 2" coarse wallboard screw at each pilot hole and into the mating ends of the end form strips.

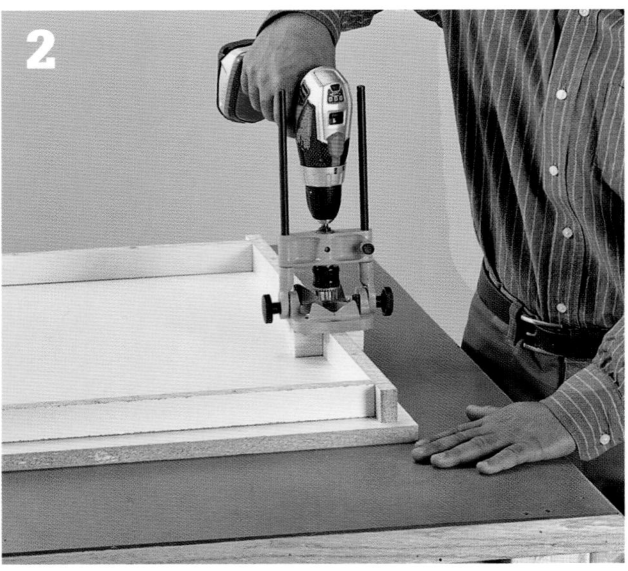

Use a power drill mounted in a right-angle drill guide (or use a drill press) to drill ¼"-dia. guide holes for 3" deck screws at 6" intervals all the way through the tops of the form sides. Countersink the holes so the screw heads will be recessed slightly below the surface.

Attach the form sides to the base. Center the melamine-strip frame pieces on the base, which should have the melamine coating face-up. Test the corners with a carpenter's square to make sure they're square. Drive one 3½" deck screw per form side near the middle. The screwheads should be slightly below the top edges of the forms. Check for square again, and continue driving the 3½" screws at 6" intervals through the pilot holes. Check for square frequently. *Note: Do not drive any screws up through the underside of the form base—you won't be able to lift the countertop and access the screws when it's time to strip off the forms.*

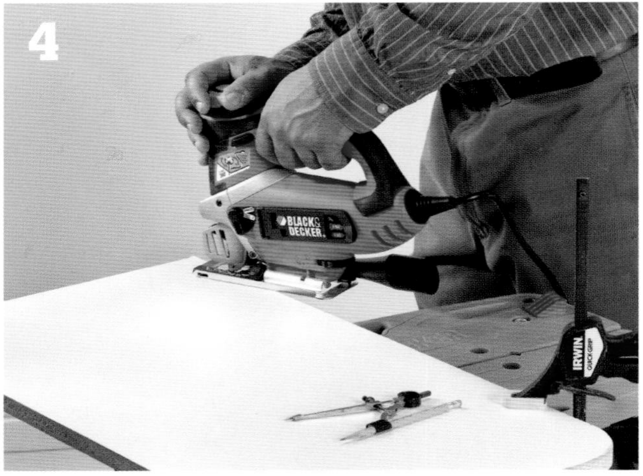

Make the sink knockout blanks by stacking two pieces of ¾" melamine. The undermount sink we used requires a 20 × 31" knockout with corners that are rounded at a 2" radius. Cut two pieces of ¾"-thick MDF to 20 × 31" square using a table saw if you have one. With a compass, mark 2"-radius curves at each corner for trimming. Make the trim cuts with a jigsaw (as shown in photo). Cut just outside the trim line and sand up to it with a pad sander for a smooth curve.

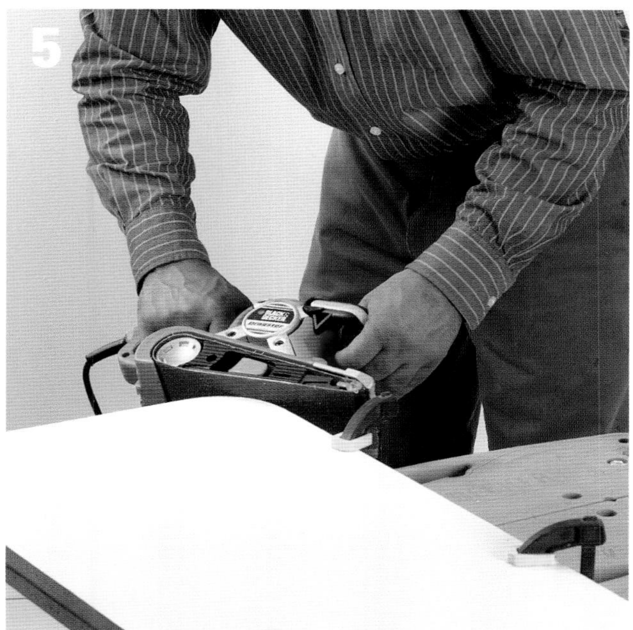

Shape the knockout. Clamp the two pieces of melamine face-to-face for the knockout and gang-sand the edges and corners so they're smooth and even. A belt sander on a stationary sanding station or an oscillating spindle sander works great for this. Don't oversand—this will cause the sink knockout to be too small.

Install the sink knockout. Because gluing the faces together can add height to the knockout (and cause the concrete finishing tools to bang into it when they ride on the form tops), attach each blank directly to the layer below it using countersunk screws. Keep the edges aligned perfectly, especially if you're planning to install an undermount sink.

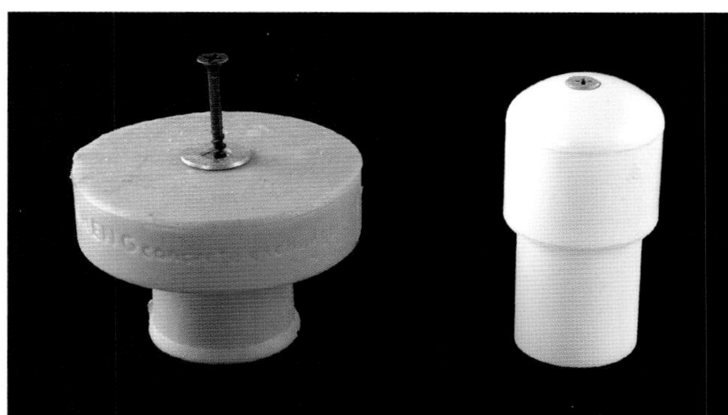

Faucet Knockouts Option: If your sink faucet will not be mounted on the sink deck, you'll need to add a knockout to your form for the faucet hole (try and choose a single-handle faucet), according to the requirements of the faucet manufacturer. You can order knockouts from a concrete countertop supplies distributor, or you can create them with PVC pipe that has an outside diameter equal to the required faucet hole size. To anchor the PVC knockout, cover one end with a flat cap made for that size tubing. Drill a guide hole through the center of the cap so you can secure it with a screw. The top of the cap should be exactly flush with the form sides once it is installed. Before securing, position the knockout next to a form side and compare the heights. If the knockout is taller, file or sand the uncapped end so their lengths match.

Make the form watertight. Seal exposed edges of the sink knockout with fast-drying polyurethane varnish, and then caulk the form once the varnish is dry. Run a very thin bead of colored silicone caulk (the coloring allows you to see where the caulk has been laid on the white melamine) in all the seams and then smooth carefully with a fingertip. In addition to keeping the wet concrete from seeping into gaps in the form, the caulk will create a slight roundover on the edges of the concrete. Caulk around the bottoms of the knockouts as well.

(continued)

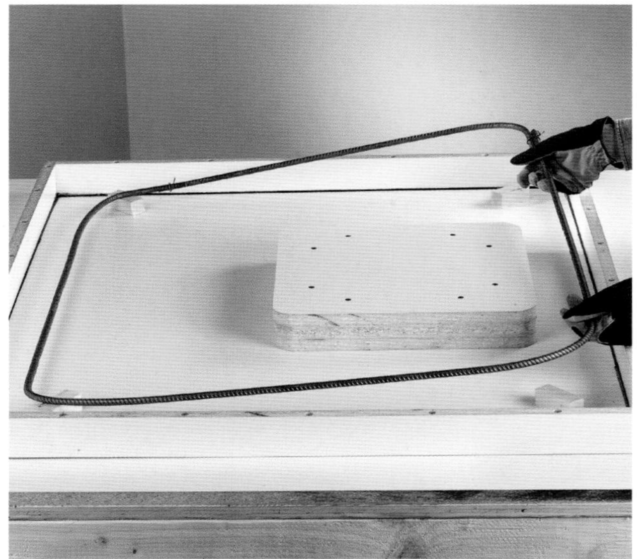

Variation: If your countertop is more than 2" thick, use #3 rebar (⅜" dia.) for the primary reinforcement. Do not use rebar on thinner countertops, as the rebar will necessarily be too close to the surface and can telegraph through. Bend the rebar to fit around the perimeter of the form using a rebar or conduit bender. The rebar needs to be at least 1" away from all edges (including knockouts) and 1" away from the top surface. Tie the ends of the rebar with wire and set it in the form on temporary 1" spacers.

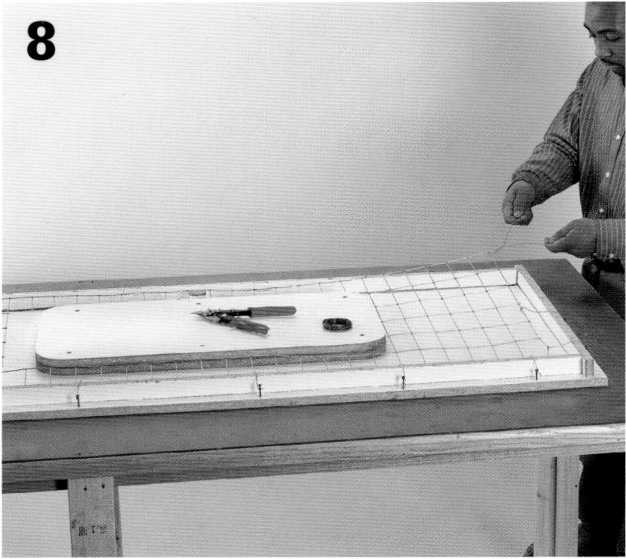

Add reinforcement. Cut a piece of welded wire (also called rewire) with a 4 × 4" grid so it's 2" smaller than the interior form dimensions. Make a cutout for the sink and faucet knockouts, making sure the rewire does not come closer than 1" to any edge, surface, or knockout. Flatten the rewire as best you can and then hang it with wires that are attached to the tops of the forms with screws (you'll remove the screws and cut the wires after the concrete is placed).

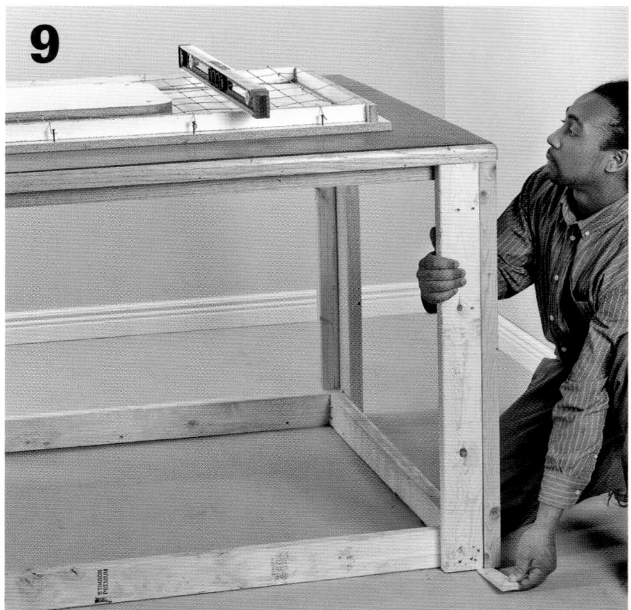

Clamp or screw the base of the form to a sturdy workbench or table so the form cannot move during the critical finishing and curing stages. Check for level and insert shims between the worktop and the benchtop if needed for leveling. If you're concerned about mess, slip a sheet of 3-mil plastic on the floor under the workbench.

Blend water with liquid cement color (if desired) in a 5-gal. bucket prior to adding to the mixer.

Slowly pour concrete countertop mix into the mixer and blend for a minimum of 5 minutes. Properly mixed material will flow easily into molds. Add small amounts of water as necessary to achieve the desired consistency.

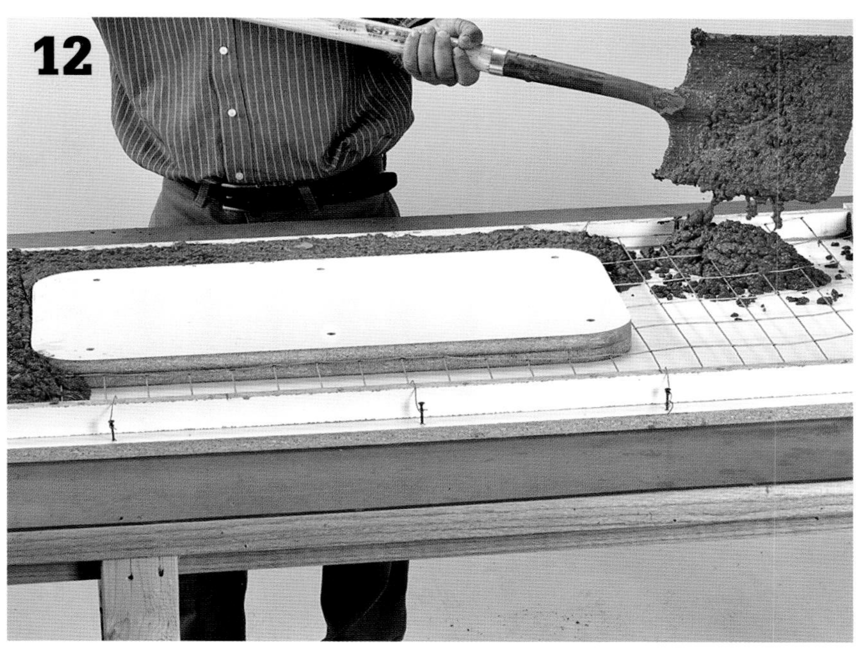

Fill the countertop form, making sure to pack the concrete into corners and press it through the reinforcement. Overfill the form slightly.

Vibrate the form vigorously as you work to settle concrete into all the voids. You can rent a concrete vibrator for this purpose, or simply strike the form repeatedly with a rubber mallet. If you have a helper and a sturdy floor and worktable, lift up and down on the ends of the table, bouncing it on the floor to cause vibrations (this is a very effective method if you can manage it safely). Make sure the table remains level when you're through.

Strike off excess concrete from the form using a 2 x 4 drawn along the tops of the forms in a sawing motion. If voids are created, pack them with fresh concrete and restrike. Do not overwork the concrete.

(continued)

Snip the wire ties holding the rewire mesh once you are certain you won't need to vibrate the form any further. Embed the cut ends attached to the rewire below the concrete surface.

Smooth the surface of the concrete with a metal screeding tool, such as a length of angle iron or square metal tubing. Work slowly with a sawing motion, allowing the bleed water to fill in behind the screed. Since this surface will be the underside of the countertop, no further tooling is required. Cover the concrete with plastic and allow the concrete to dry undisturbed for three to five days.

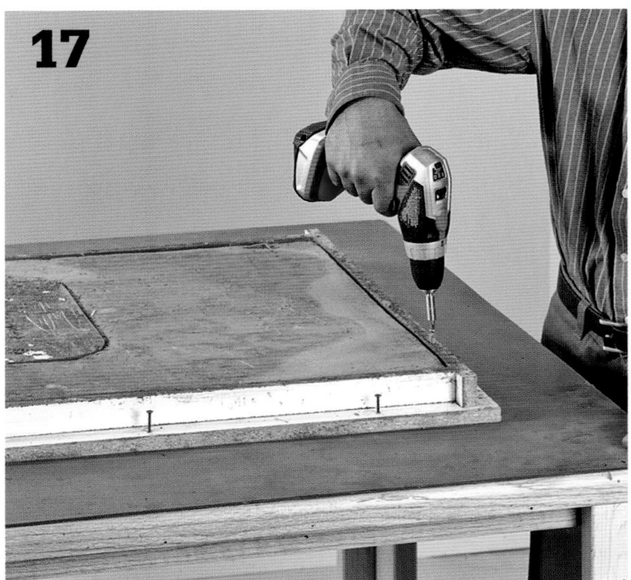

Remove the plastic covering and then unscrew and remove the forms. Do not pry against the fresh concrete. In most cases, you'll need to cut apart the sink knockout to prevent damaging the countertop when removing it. Drill a starter hole and then carefully cut up to the edge of the knockout. Cut the knockout into chunks until you can remove it all. The edges of the concrete will be fragile, so be very careful.

Flip the countertop so the finished surface is exposed (you'll need a helper or two). Be extremely careful. The best technique is to roll the countertop onto an edge, position several shock-absorbing sleepers beneath it (rigid insulation board works very well), and then gently lower the countertop onto the sleepers.

To expose the aggregate and create a very polished finish, grind the countertop surface. Use a series of increasingly fine grinding pads mounted on a shock-protected 5" angle grinder (variable speed). This is messy work and can go on for hours to get the desired result. Rinse the surface regularly with clean water and make sure it stays wet during grinding. For a gleaming surface, mount still finer pads (up to 1,500 grit) on the grinder and wet-polish.

Clean and seal the concrete with several coats of quality concrete sealer (one with penetrating and film-forming agents). For extra protection and a renewable finish, apply a coat of paste wax after the last coat of sealer dries.

Mount the sink (if undermount). Sinks are easier to install prior to attaching the countertop on the cabinet. Attach the sink according to the manufacturer's directions. Undermount sinks like this are installed with undermount clips and silicone adhesive. Self-rimming sinks likely will require some modifications to the mounting hardware (or at least you'll need to buy some extra-long screws) to accommodate the thickness of the countertop.

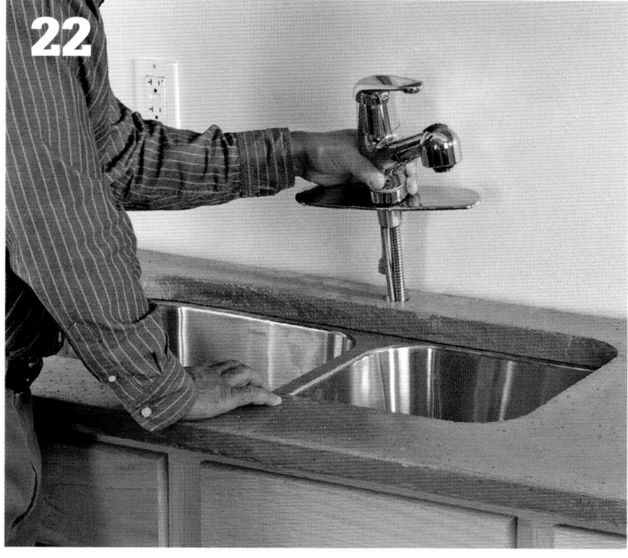

Install the countertop and hook up the plumbing. Make sure the island cabinet is adequately reinforced and that as much plumbing as possible has been taken care of, and then apply a thick bead of panel adhesive or silicone adhesive to the tops of the cabinets and stretchers. With at least one helper, lower the countertop onto the base and position it where you wish. Let the adhesive dry overnight before completing the sink and faucet hookups.

Brick & Block

Brick and block add an earthen, structural feel to an indoor or outdoor environment. The material spans many applications and the array of block material available today in different textures, colors, and sizes allows you to customize the projects in this chapter to your liking. If you are just learning masonry and stonework skills, a brick and block project is a great place to start.

Most brick and block projects are permanently bound together with mortar, including setting the first course into a bed of mortar on a stable footing. In some cases, however, the masonry units can be dry-laid (stacked without using mortar) and then bound together with a coat of stucco or masonry veneer that is applied to the surfaces after the stacking is completed. With the exception of a dry-stacked stone wall, this kid of wall represents the easiest starter masonry project. Be certain the base is stable, however, as any movement will cause visible cracking and eventual failure if repairs are not enacted quickly.

In this chapter:

Brick & Block Basics

Laying brick and block is a precise business. Many of the tools necessary for these projects relate to establishing and maintaining true, square and level structures, while others relate to cutting the masonry units and placing the mortar. It makes sense to purchase tools you'll use again, but it's more cost effective to rent specialty items, such as a brick splitter.

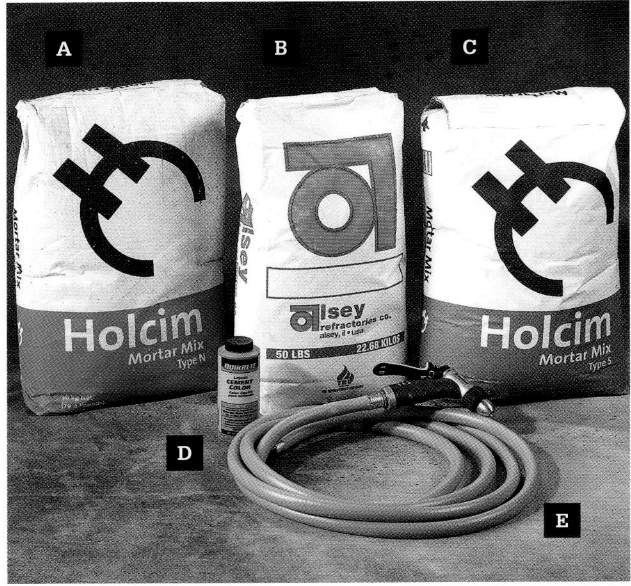

Mortar mixes: (A) Type N, a medium-strength mortar for above-grade outdoor use in nonload-bearing (freestanding) walls, barbeques, chimneys, and tuck-pointing; (B) refractory mortar, a calcium aluminate mortar that is resistant to high temperatures, used for mortaring around firebrick in fireplaces and barbeques; (C) Type S, a high-strength mortar for outdoor use at or below grade, typically used in foundations, retaining walls, driveways, walks, and patios; (D) mortar tint for coloring mortar; (E) and you'll need water for mixing mortar so a hose is needed (a sprayer attachment is needed later to clean surface).

Common types of brick and block used for residential construction include: decorative block (A) available colored or plain; decorative concrete pavers (B); fire brick (C); standard 8 × 8 × 16" concrete block (D); half block (E); combination corner block (F); queen-sized brick (G); standard brick pavers (H); standard building bricks (I); and limestone wall cap (J).

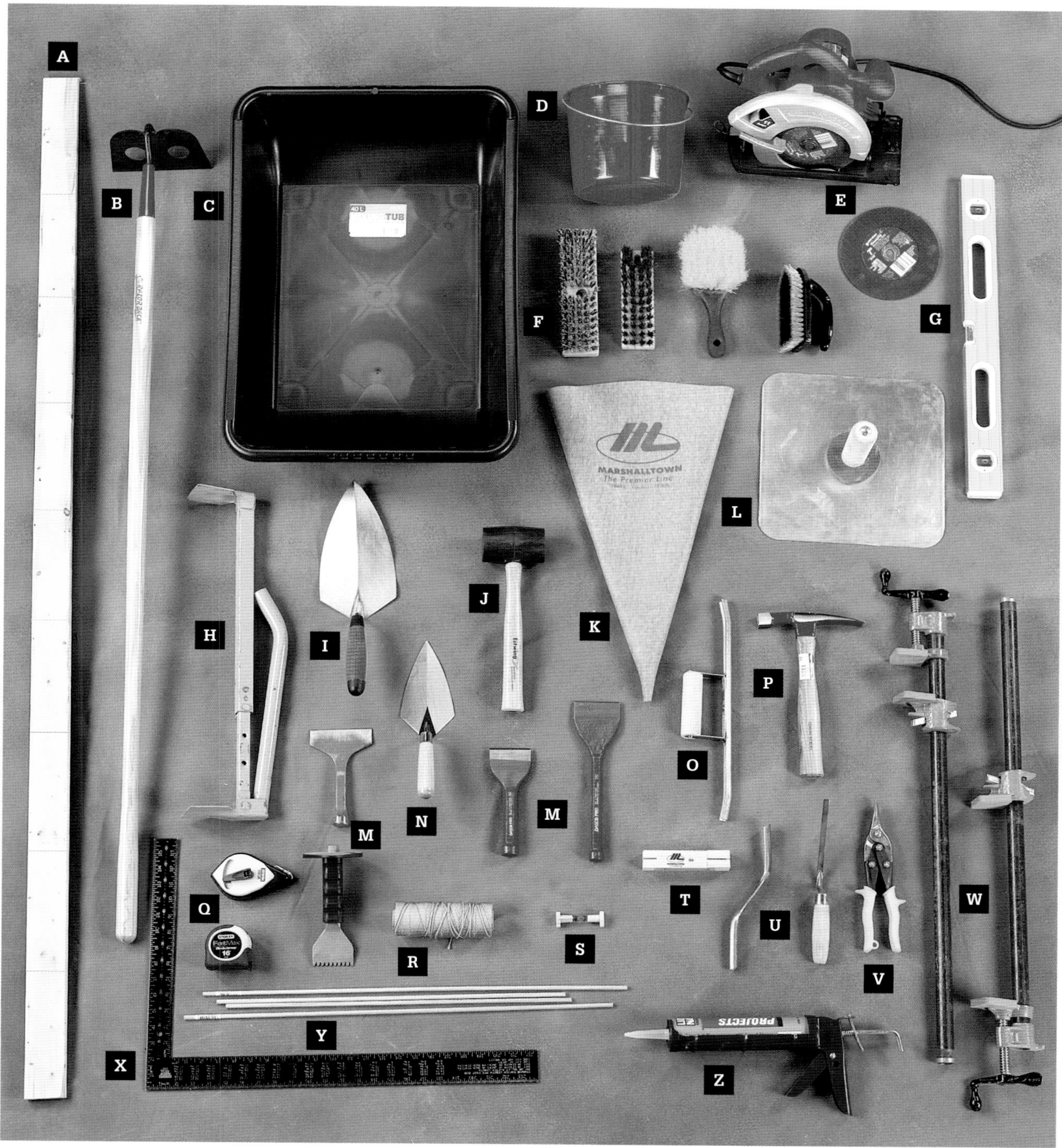

Mason's tools include: a story pole (A) for checking stacked masonry units; masonry hoe (B) and mortar box (C) for mixing mortar; a bucket (D) and stiff bristle brushes (F) for removing stains and loose materials; circular saw and masonry-cutting blades (E) for scoring brick and block; level (G) for checking stacked masonry units; brick tongs (H) for carrying multiple bricks; mason's trowel (I) for applying mortar; rubber mallet (J) for setting paver stones; mortar bag (K) for filling horizontal joints; mortar hawk (L) for holding mortar; masonry chisels (M) for splitting brick, block, and stone; pointing trowel (N) for placing mortar on brick and block walls; sled jointer (O) for finishing long joints; mason's hammer (P) for chipping brick and stone; a tape measure and chalk line (Q) for marking layout lines on footings or slabs; mason's string (R) and line blocks (T) for stacking brick and block; a line level (S) for making layouts and setting slope; jointers (U) for finishing mortar joints; aviation snips (V) for trimming metal ties and lath; pipe clamps (W) for aligning brick and block to be scored; a framing square (X) for setting project outlines; ⅜" dowels (Y) for spacers between dry-laid masonry units; caulk gun (Z) for sealing around fasteners and house trim.

Planning & Techniques

Like other masonry projects, brick and block projects must start with careful planning. You need to identify the construction techniques and methods that are appropriate for the project, practice any techniques you need to learn, and estimate and order your materials.

Estimating Bricks & Blocks ▸

Standard brick pavers for walks and patios (4 × 8)	surface area (sq. ft.) × 5 = number of pavers needed
Standard bricks for walls and pillars (4 × 8)	surface area (sq. ft.) × 7 = number of pavers needed (single brick thickness)
Interlocking block	area of wall face (sq. ft.) × 1.5 = number of blocks needed
8 × 8 × 16 concrete for freestanding walls	Height of wall (ft.) × length of wall × 1.125 = number of blocks needed

Select a construction design that makes sense for your project. There are two basic methods used in stacking brick or block. Structures that are only one unit wide are called single wythe and are typically used for projects like brick barbecues or planters, and for brick veneers. Double-wythe walls are two units wide and are used in free-standing applications. Most concrete-block structures are single wythe.

Keep structures as low as you can. Local codes require frost footings and additional reinforcement for permanent walls or structures that exceed maximum height restrictions. You can often simplify your project by designing walls that are below the maximum height.

Add a lattice panel or another decorative element to permanent walls to create greater privacy without having to add structural reinforcement to the masonry structure.

How to Plan a Brick or Block Project

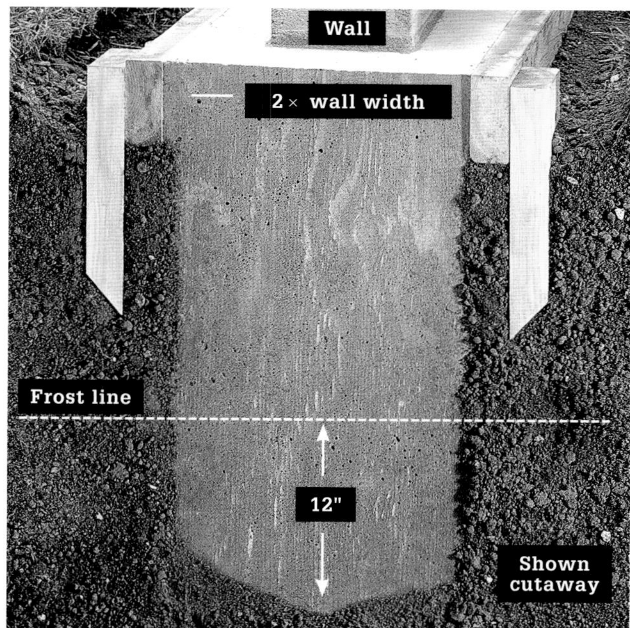

Frost footings are required if a structure will be more than 2 ft. tall or if it is tied to another permanent structure. Frost footings should be twice as wide as the structure they support and should extend 8 to 12" below the frost line (pages 46 to 47).

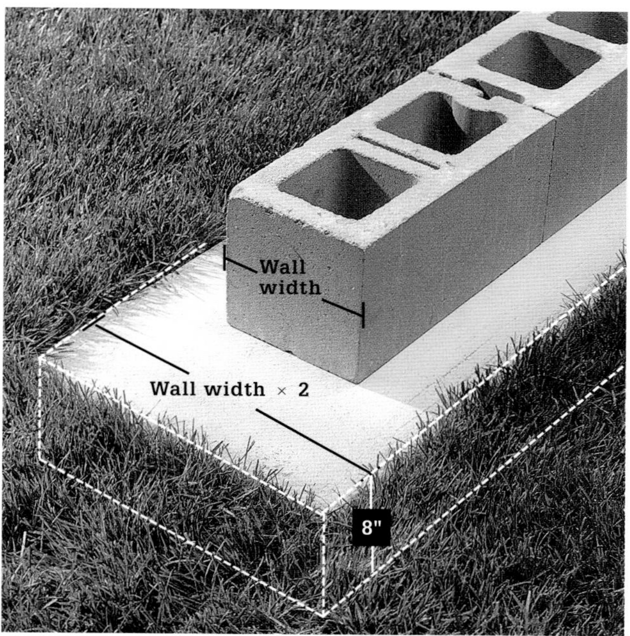

Pour a reinforced concrete slab for brick and block structures that are freestanding and under 2 ft. tall. The slab should be twice as wide as the wall, flush with ground level, and at least 8" thick. Check with building codes for special requirements. Slabs are poured using the techniques for pouring a sidewalk (pages 50 to 55).

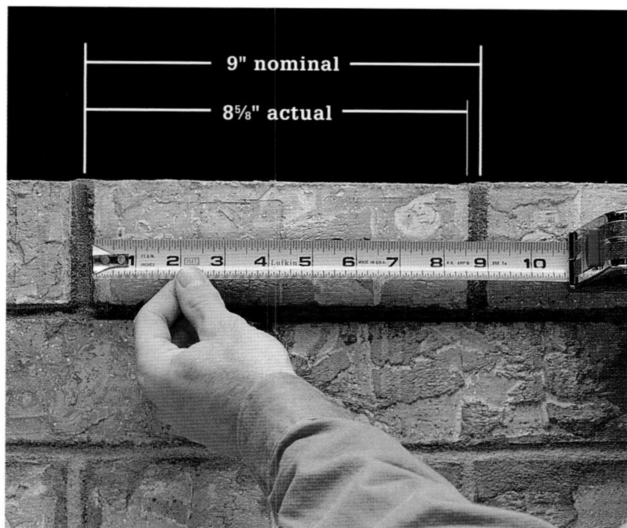

Do not add mortar joint thickness to total project dimensions when planning brick and block projects. The actual sizes of bricks and blocks are ⅜" smaller than the nominal size to allow for ⅜"-wide mortar joints. For example, a 9" (nominal) brick has an actual dimension of 8⅝", so a wall that is built with four 9" bricks and ⅜" mortar joints will have a finished length of 36" (4 × 9").

Test project layouts using ⅜" spacers between masonry units to make sure the planned dimensions work. If possible, create a plan that uses whole bricks or blocks, reducing the amount of cutting required.

Reinforcing Brick & Block Structures

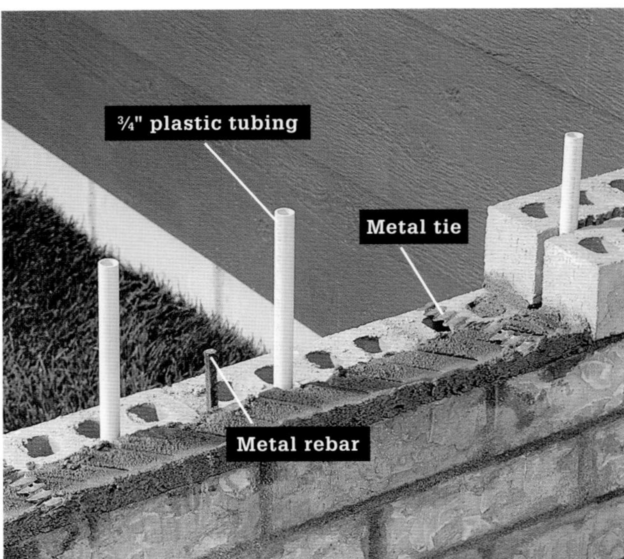

For double-wythe brick projects, use metal ties between wythes for reinforcement. Insert ties directly into the mortar 2 to 3 ft. apart, every third course. Insert metal rebar into the gap between wythes every 4 to 6 ft. (check local building codes). Insert ¾"-diameter plastic tubing between wythes to keep them aligned. Pour a thin mixture of mortar between the wythes to improve the strength of the wall.

For block projects, fill the empty spaces (cores) of the block with thin mortar. Insert sections of metal rebar into the mortar to increase vertical strength. Check with your local building inspector to determine reinforcement requirements, if any.

Provide horizontal reinforcement on brick or block walls by setting metal reinforcing strips into the mortar every third course. Metal reinforcing strips, along with most other reinforcing products, can be purchased from brick and block suppliers. Overlap the ends of metal strips 6" where they meet.

Tips for Working with Brick ▸

Make practice runs on a 2 × 4 to help you perfect your mortar-throwing (pages 118 to 119) and bricklaying techniques. You can clean and reuse the bricks to make many practice runs if you find it helpful, but do not reuse the bricks in your actual project—old mortar can impede bonding.

Test the water absorption rate of bricks to determine their density. Squeeze out 20 drops of water in the same spot on the surface of a brick. If the surface is completely dry after 60 seconds, dampen the bricks with water before you lay them to prevent them from absorbing moisture from the mortar before it has a chance to set.

Use a T-square and pencil to mark several bricks for cutting. Make sure the ends of the bricks are all aligned.

Mark angled cuts by dry-laying the project (as shown with pavers above) and setting the brick or block in position. Allow for ⅜" joints in mortared projects. Pavers have spacing lugs that set the spacing at ⅛". Mark cutting lines with a pencil using a straightedge where practical to mark straight lines.

How to Score & Cut Brick

Score all four sides of the brick first with a brickset chisel and maul when cuts fall over the web area and not over the core. Tap the chisel to leave scored cutting marks ⅛ to ¼" deep, and then strike a firm final blow to the chisel to split the brick. Properly scored bricks split cleanly with one firm blow.

Option: When you need to split a lot of bricks uniformly and quickly, use a circular saw fitted with a masonry blade to score the bricks, then split them individually with a chisel. For quick scoring, clamp them securely at each end with a pipe or bar clamp, making sure the ends are aligned. Remember: Wear eye protection when using striking or cutting tools.

How to Angle-cut Brick

Mark the final cutting line on the brick. To avoid ruining the brick, you will need to make gradual cuts until you reach this line. Score a straight line for the first cut in the waste area of the brick about ⅛" from the starting point of the final cutting line, perpendicular to the edge of the brick. Make the first cut.

Keep the chisel stationary at the point of the first cut, pivot it slightly, and then score and cut again. It is important to keep the pivot point of the chisel at the edge of the brick. Repeat until all of the waste area is removed.

How to Use a Brick Splitter

A brick splitter makes accurate, consistent cuts in bricks and pavers with no scoring required. It is a good idea to rent one if your project requires many cuts. To use the brick splitter, first mark a cutting line on the brick, and then set the brick on the table of the splitter, aligning the cutting line with the cutting blade on the tool.

Once the brick is in position on the splitter table, pull down sharply on the handle. The cutting blade on the splitter will cleave the brick along the cutting line. *Tip: For efficiency, mark cutting lines on several bricks at the same time.*

How to Cut Concrete Block

Mark cutting lines on both faces of the block, and then score ⅛ to ¼"-deep cuts along the lines using a circular saw equipped with a masonry blade.

Use a mason's chisel and maul to split one face of the block along the cutting line. Turn the block over and split the other face.

Option: Cut half blocks from combination corner blocks. Corner blocks have preformed cores in the center of the web. Score lightly above the core, and then rap with a mason's chisel to break off half blocks.

Mixing & Placing Mortar

A professional bricklayer at work is an impressive sight, even for do-it-yourselfers who have completed numerous masonry projects successfully. The mortar practically flies off the trowel and seems to end up in perfect position to accept the next brick or block.

Although "throwing mortar" is an acquired skill that takes years to perfect, you can use the basic techniques successfully with just a little practice.

The first critical element to handling mortar effectively is the mixture. If it's too thick, it will fall off the trowel in a heap, not in the smooth line that is your goal. Add too much water and the mortar becomes messy and weak. Follow the manufacturer's directions, but keep in mind that the amount of water specified is an approximation. If you've never mixed mortar before, experiment with small amounts until you find a mixture that clings to the trowel just long enough for you to deliver a controlled, even line that holds its shape after settling. Note how much water you use in each batch, and record the best mixture.

Mix mortar for a large project in batches; on a hot, dry day a large batch will harden before you know it. If mortar begins to thicken, add water (called retempering); use retempered mortar within two hours.

Tools & Materials ▸

Mortar mix	Mason's trowel
Mortar box	Bricks
Masonry hoe	Mortar tint
Plywood	Work gloves

How to Mix & Place Mortar

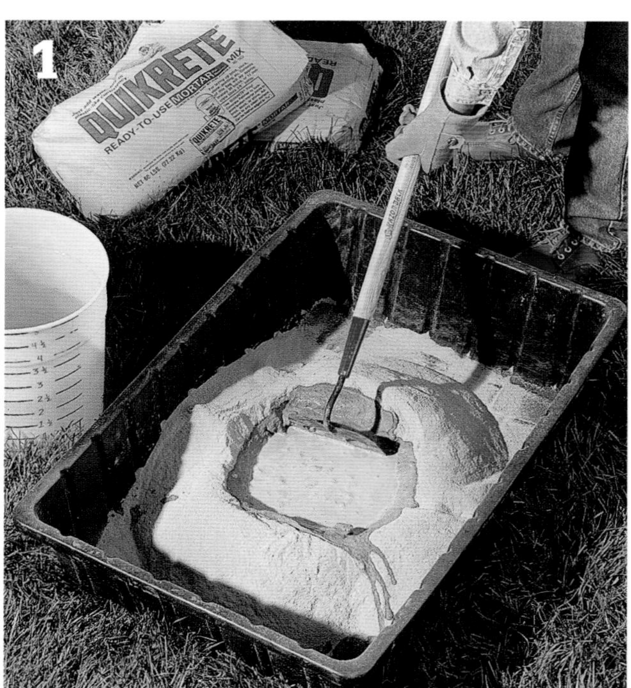

Empty mortar mix into a mortar box and form a depression in the center. Add about ¾ of the recommended amount of water into the depression, and then mix it in with a masonry hoe. Do not overwork the mortar. Continue adding small amounts of water and mixing until the mortar reaches the proper consistency. Do not mix too much mortar at one time—mortar is much easier to work with when it is fresh.

Set a piece of plywood on blocks at a convenient height, and place a shovelful of mortar onto the surface. Slice off a strip of mortar from the pile using the edge of your mason's trowel. Slip the trowel point-first under the section of mortar and lift up.

Snap the trowel gently downward to dislodge excess mortar clinging to the edges. Position the trowel at the starting point, and "throw" a line of mortar onto the building surface. A good amount is enough to set three bricks. Do not get ahead of yourself. If you throw too much mortar, it will set before you are ready.

"Furrow" the mortar line by dragging the point of the trowel through the center of the mortar line in a slight back-and-forth motion. Furrowing helps distribute the mortar evenly.

Mortar Mixing Tips ▸

Adding tint to mortar works best if you add the same amount to each batch throughout the project. Once you settle on a recipe, record it so you can mix the same proportions each time.

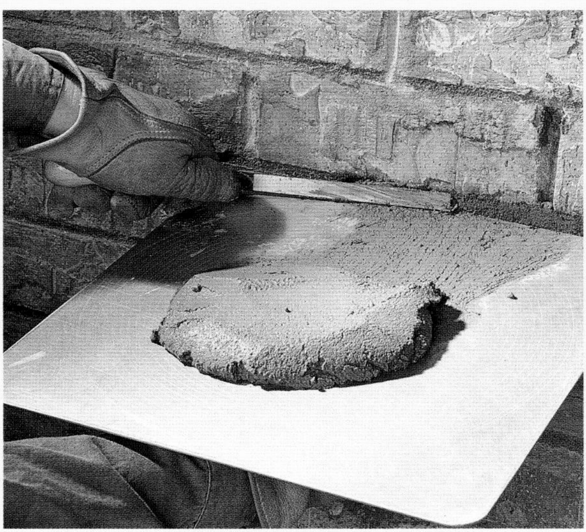

Use a stiff (dry) mix of mortar for tuck-pointing—it's less likely to shrink and crack. Start by mixing Type N mortar mix with half the recommended water. Let the mixture stand for one hour, and then add the remaining water and finish mixing.

Layering Brick

Patience, care, and good technique are the key elements to building brick structures that have a professional look. Start with a sturdy, level footing (pages 46 to 49), and don't worry if your initial bricklaying attempts aren't perfect. Survey your work often and stop when you spot a problem. As long as the mortar's still soft, you can remove bricks and try again.

This section features one method of brick wall construction: laying up the ends of the wall first, and then filling in the interior bricks. The alternate method, laying one course at a time, is shown with concrete block (pages 124 to 127).

Tools & Materials ▸

- Gloves
- Trowel
- Chalk line
- Level
- Line blocks
- Mason's string
- ⅜" dowels
- Jointing tool
- Mortar
- Brick
- Wall ties
- Rebar (optional)
- Eye protection

Buttering is a term used to describe the process of applying mortar to the end of a brick or block before adding it to the structure being built. Apply a heavy layer of mortar to one end of a brick, and then cut off the excess with a trowel.

How to Build a Double-layered Brick Wall

Dry-lay the first course by setting down two parallel rows of brick, spaced ¾ to 1" apart. Use a chalk line to outline the location of the wall on the slab. Draw pencil lines on the slab to mark the ends of the bricks. Test-fit the spacing with a ⅜"-dia. dowel, and then mark the locations of the joint gaps to use as a reference after the spacers are removed.

Dampen the concrete slab or footing with water, and dampen the bricks or blocks if necessary. Mix mortar and throw a layer of mortar on to the footing for the first two bricks of one wythe at one end of the layout. Butter the inside end of the first brick, and then press the brick into the mortar, creating a ⅜" mortar bed. Cut away excess mortar.

Plumb the face of the end brick using a level. Tap lightly with the handle of the trowel to correct the brick if it is not plumb. Level the brick end to end. Butter the end of a second brick, and then set it into the mortar bed, pushing the dry end toward the first brick to create a joint of ⅜".

Butter and place a third brick using the chalk lines as a general reference, and then using a level to check for level and plumb. Adjust any bricks that are not aligned by tapping lightly with the trowel handle.

Lay the first three bricks for the other wythe, parallel to the first wythe. Level the wythes, and make sure the end bricks and mortar joints align. Fill the gaps between the wythes at each end with mortar.

Cut a half brick, and then throw and furrow a mortar bed for a half brick on top of the first course. Butter the end of the half brick, and then set the half brick in the mortar bed, creating a ⅜" joint. Cut away excess mortar. Make sure bricks are plumb and level.

Add more bricks and half bricks to both wythes at the end until you lay the first bricks in the fourth course. Align bricks with the reference lines. *Note: To build corners, lay a header brick at the end of two parallel wythes. Position the header brick in each subsequent course perpendicular to the header brick in the previous course (inset).*

(continued)

Check the spacing of the end bricks with a straightedge. Properly spaced bricks will form a straight line when you place the straightedge over the stepped end bricks. If bricks are not in alignment, do not move those bricks already set. Try to compensate for the problem gradually as you fill in the middle (field) bricks (Step 11) by slightly reducing or increasing the spacing between the joints.

Every 30 minutes, stop laying bricks and smooth out all the untooled mortar joints with a jointing tool. Do the horizontal joints first, and then the vertical joints. Cut away any excess mortar pressed from the joints using a trowel. When the mortar has set, but is not too hard, brush any excess mortar from the brick faces.

Build the opposite end of the wall with the same methods as the first using the chalk lines as a reference. Stretch a mason's string between the two ends to establish a flush, level line between ends—use line blocks to secure the string. Tighten the string until it is taut. Begin to fill in the field bricks (the bricks between ends) on the first course using the mason's string as a guide.

Lay the remaining field bricks. The last brick, called the closure brick, should be buttered at both ends. Center the closure brick between the two adjoining bricks, and then set in place with the trowel handle. Fill in the first three courses of each wythe, moving the mason's string up one course after completing each course.

In the fourth course, set metal wall ties into the mortar bed of one wythe and on top of the brick adjacent to it. Space the ties 2 to 3 ft. apart, every three courses. For added strength, set metal rebar into the cavities between the wythes and fill with thin mortar.

Lay the remaining courses, installing metal ties every third course. Check with mason's string frequently for alignment, and use a level to make sure the wall is plumb and level.

Lay a furrowed mortar bed on the top course, and place a wall cap on top of the wall to cover empty spaces and provide a finished appearance. Remove any excess mortar. Make sure the cap blocks are aligned and level. Fill the joints between cap blocks with mortar.

Laying Block

Block walls can be built fairly quickly because of the size of the individual blocks. Still, the same patience and attention to detail involved in laying bricks is required. Check your work often, and don't be afraid to back up a step or two to correct your mistakes.

This section features a concrete block wall laid up one course at a time. Make sure you have a sturdy, level footing (page 46 to 49) before you start.

Tools & Materials ▸

- Trowel
- Chalk line
- Level
- Mason's string
- Line blocks
- Jointing tool
- Stakes
- Work gloves
- Steel truss
- Work braces
- Mortar mix
- 8 × 8" concrete blocks
- Stakes
- Cap blocks
- Rebar
- Wire reinforcing strips
- Scrap lumber
- Hammer
- Chisel

Buttering a concrete block involves laying narrow slices of mortar on the two flanges at the end of the block. It is not necessary to butter the valley between the flanges unless the project calls for it.

How to Lay a Concrete Block Wall

Dry-lay the first course, leaving a ⅜" gap between blocks. Draw reference lines on the concrete base to mark the ends of the row, extending the lines well past the edges of the block. Use a chalk line to snap reference lines on each side of the base, 3" from the blocks. These reference lines will serve as a guide when setting the blocks into mortar.

Dampen the base slightly, then mix mortar, and throw and furrow two mortar lines at one end to create a mortar bed for the combination corner block. Dampen porous blocks before setting them into the mortar beds.

Set a combination corner block (page 110) into the mortar bed. Press it into the mortar to create a ⅜"-thick bed joint. Hold the block in place and cut away the excess mortar (save excess mortar for the next section of the mortar bed). Check the block with a level to make sure it is level and plumb. Make any necessary adjustments by rapping on the high side with the handle of a trowel. Be careful not to displace too much mortar.

Drive a stake at each end of the project and attach one end of a mason's string to each stake. Thread a line level onto the string and adjust the string until it is level and flush with the top of the corner block. Throw a mortar bed and set a corner block at the other end. Adjust the block so it is plumb and level, making sure it is aligned with the mason's string.

Throw a mortar bed for the second block at one end of the project: Butter one end of a standard block and set it next to the corner block, pressing the two blocks together so the joint between them is ⅜" thick. Tap the block with the handle of a trowel to set it and adjust the block until it is even with the mason's string. Be careful to maintain the ⅜" joint.

Install all but the last block in the first course, working from the ends toward the middle. Align the blocks with the mason's string. Clean excess mortar from the base before it hardens.

(continued)

Butter the flanges on both ends of a standard block for use as the closure block in the course. Slide the closure block into the gap between blocks, keeping the mortar joints an even thickness on each side. Align the block with the mason's string.

Apply a 1"-thick mortar bed for the half block at one end of the wall, and then begin the second course with a half block.

Set the half block into the mortar bed with the smooth surfaces facing out. Use the level to make sure the half block is plumb with the first corner block, and then check to make sure it is level. Adjust as needed. Install a half block at the other end.

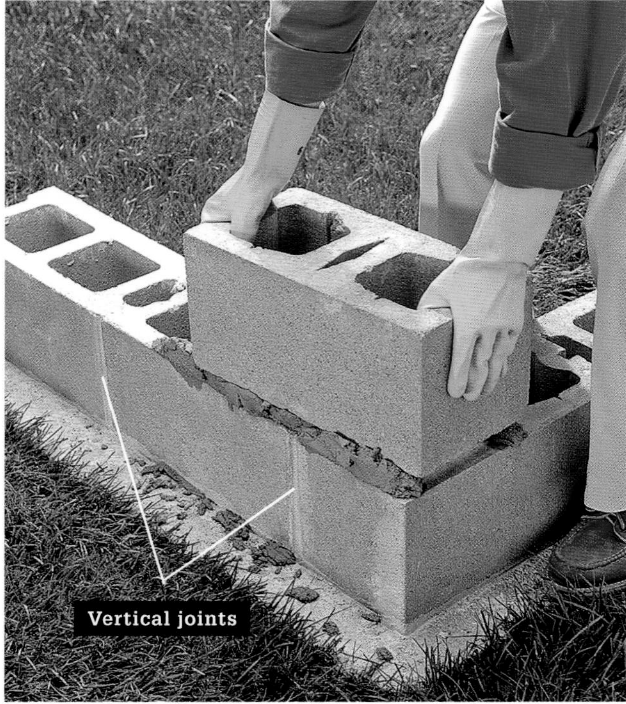

Variation: If your wall has a corner, begin the second course with a full-sized end block that spans the vertical joint formed where the two walls meet. This layout creates and maintains a running bond for the wall.

Attach a mason's string for reference, securing it either with line blocks or a nail. If you do not have line blocks, insert a nail into the wet mortar at each end of the wall, and then wind the mason's string around and up to the top corner of the second course, as shown above. Connect both ends and draw the mason's string taut. Throw a mortar bed for the next block, and then fill out the second course using the mason's string as a reference line.

Every half-hour, tool the fresh mortar joints with a jointing tool and remove any excess mortar. Tool the horizontal joints first, and then the vertical joints. Cut off excess mortar using a trowel blade. When the mortar has set, but is not too hard, brush any excess mortar from the block faces. Continue building the wall until it is complete.

Option: When building stack bond walls with vertical joints that are in alignment, use wire reinforcing strips in the mortar beds every third course (or as required by local codes) to increase the strength of the wall. The wire should be completely embedded in the mortar. See page 114 for other block wall reinforcing options.

Install a wall cap on top of the wall to cover the empty spaces and create a finished appearance. Set the cap pieces into mortar beds, and then butter an end with mortar. Level the cap, and then tool to match the joints in the rest of the wall.

How to Build a Foundation Wall

Position story poles at each corner of the foundation. Mark the top line using the string lines as reference, and then mark down 8" for each course of blocks.

Lay the first course in a "dry run" to determine if you'll need to cut or use any special blocks. Use a scrap piece ⅜" thick as a spacer for the mortar joints.

Lay first block by spreading enough mortar for three blocks in a ladder pattern. Set corner block in place and check plumb and level against story pole.

Set the opposite corner block in place and position mason line blocks and guide string. Follow the string in laying the rest of the course.

As you build courses, tie supporting rebar to the footing rebar. Fill the rebar cavities with concrete.

Cut blocks as necessary, scoring on mark and breaking with hammer and broad, cold chisel. You can also use a grinder equipped with a diamond blade.

Tool the mortar joints with a mason's jointer when the mortar is "thumbprint" ready. Sink anchor bolts for the mudsill into block cavities filled with concrete (inset). Space them roughly every 4 ft.

Provide additional lateral reinforcement by using steel truss work braces or "ladders" made to lay across the top of a course around the block cavities. These should be used every other course for best effect. Lastly, you can use special metal lath to isolate a course and fill just that course with concrete—a requirement of some codes for the top-most course.

Dry Block Wall

The project shown here demonstrates how to lay a mortarless concrete block wall that is coated with surface-bonding cement. Surface-bonding cement contains thousands of fiberglass fibers that interlock when the product cures, giving the wall greater flexural strength than an ordinary mortared block wall. The finished appearance of walls coated with surface-bonding cement resembles stucco. It is ideal for garden walls, stucco fence walls, trash can enclosures, mobile home skirting, and as a waterproof coating for concrete ponds when used with acrylic fortifier.

Tools & Materials ▸

- Aviation snips
- Mason's trowel
- Brickset
- Chisel
- Maul
- Mason's string
- Level
- Chalk line
- Circular saw
- Sprayer
- Mortar
- Mortar box
- Line blocks
- Concrete block
- Metal ties
- Wire mesh
- Surface-bonding cement
- Stucco and mortar color (optional)
- Eye protection and work gloves

Surface bonding cement gives a dry-stacked block wall an attractive finished appearance. It also binds the blocks together.

Moisten the blocks before applying the product.

How to Lay a Mortarless Block Wall

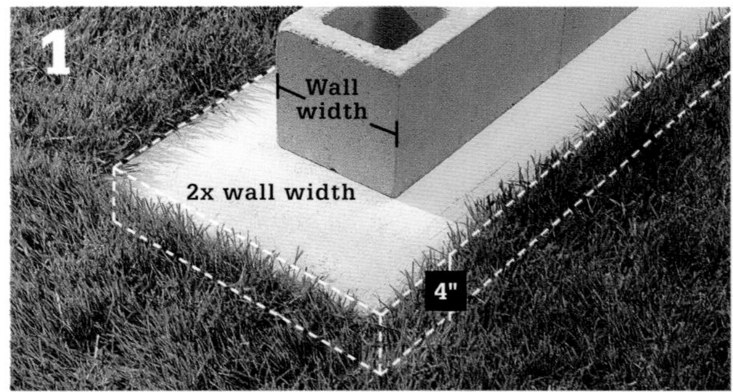

Start with a dry layout of the first course on a concrete footing. Where less than half a block is needed, trim two blocks instead. For example, where three and one-third block lengths are required, use four blocks, and cut two of them to two-thirds their length. You'll end up with a stronger, more durable wall.

Mark the corners of the end blocks on the footing with a pencil. Then, remove the blocks and snap chalk lines to indicate where to lay the mortar bed and the initial course of block.

Mist the footing with water, and then lay a ¾"-thick bed of mortar on the footing. Take care to cover only the area inside the reference lines. The mortar must be firm enough to prevent the first course from sagging.

Lay the first course, starting at one end and placing blocks in the mortar bed with no spacing in between. Use solid-faced blocks on the ends of the wall and check the course for level. If your wall is longer than 20 ft., consider inclusion of an expansion joint.

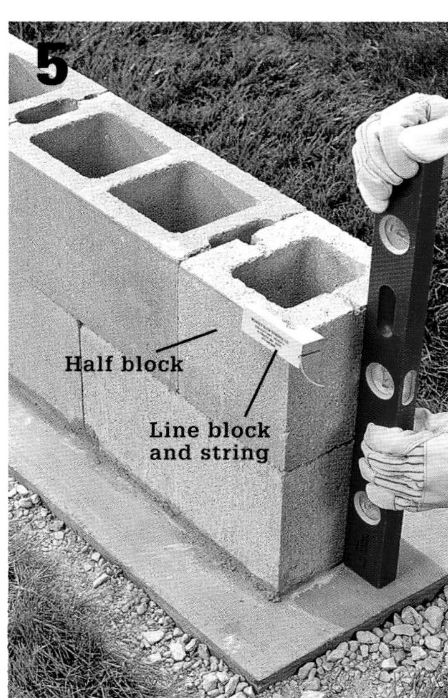

Lay subsequent courses one at a time using a level to check for plumb and line blocks to check for level. Begin courses with solid-faced blocks at each end. Use half blocks to establish a running-bond pattern. For walls over 6 ft. tall, consult local building codes.

Mix the surface-bonding cement thoroughly in a mortar box until it achieves a firm, workable consistency. Eliminate all lumps during mixing. If you are coloring the surface bonding cement, add the coloring agent directly to the mixing water prior to mixing the product.

Apply surface-bonding cement in a ¼"-thick layer. Work from the bottom of the wall to the top. A variety of stucco textures can be added to the wall as soon as it becomes thumbprint hard.

Block Retaining Wall

Retaining walls are often used to level a yard or to prevent erosion on a hillside. In a flat yard, you can build a low retaining wall and fill in behind it to create a raised planting bed.

While retaining walls can be built from many materials, such as pressure-treated timbers and natural stone, interlocking blocks are common. Typically made from concrete, interlocking retaining wall blocks are rather inexpensive, very durable, and DIY friendly. Several styles of interlocking block are available at building centers and landscape materials suppliers. Most types have a natural rock finish that combines the rough texture of cut stone with the uniform shape and size of concrete blocks.

Interlocking block weighs up to 80 pounds each, so it is a good idea to have helpers when building a retaining wall. Suppliers offer substantial discounts when interlocking block is purchased in large quantities, so you may be able to save money if you coordinate your own project with those of your neighbors.

The retaining walls in this section were built with either interlocking block or cut stone. These durable materials are easy to work with. No matter what material you use, your wall can be damaged if water saturates the soil behind it, so make sure you include the proper drainage features. You may need to dig a drainage swale before building in low-lying areas.

Tools & Materials ▸

Wheelbarrow
Shovel
Garden rake
Line level
Hand tamper
Tamping machine
Small maul
Masonry chisel
Eye protection
Hearing protectors
Work gloves
Circular saw with masonry-cutting blade
Level
Tape measure
Marking pencil
Caulk gun
Stakes
Mason's string
Landscape fabric
Compactable gravel
Perforated drain pipe
Coarse backfill material
Construction adhesive
Retaining wall block
Cap blocks
Spraypaint

Terraced retaining walls work well on steep hillsides. Two or more short retaining walls are easier to install and more stable than a single, tall retaining wall. Construct the terraces so each wall is no higher than 3 ft.

Options for Positioning a Retaining Wall

You can add more level ground to your yard by how you position a retaining wall. To increase the level area above the wall, position it well forward from the top of the hill. Fill in behind the wall with extra soil. Maintain the shape of your yard by positioning the wall near the top of the hillside, and use soil removed at the base of the hill to fill in behind the top of the wall.

Structural Features ▸

The "guts" of a retaining wall are its structural features: a compactable gravel sub-base to make a solid footing for the wall; crushed stone backfill and a perforated drain pipe to improve drainage behind the wall; and landscape fabric to keep the loose soil from washing into and clogging the gravel backfill.

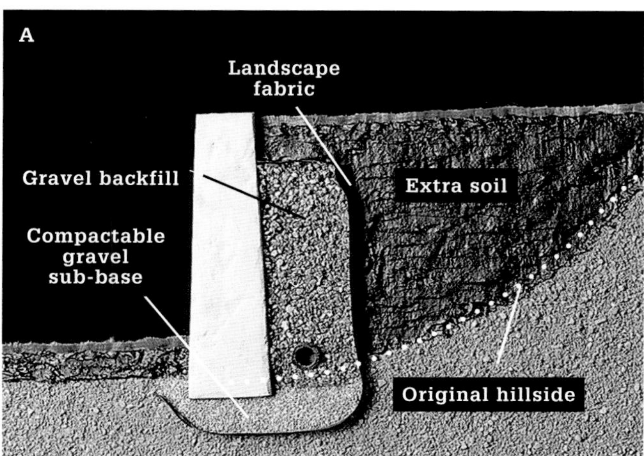

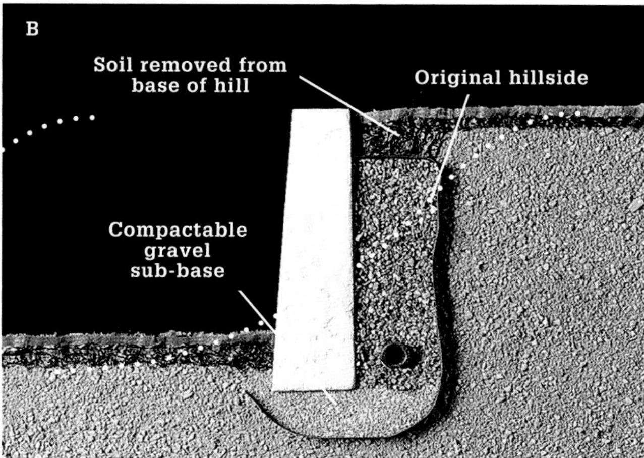

Increase the level area above the wall (A) by positioning the wall well forward from the top of the hill. Fill in behind the wall with extra soil, which is available from sand-and-gravel companies. **Keep the basic shape** of your yard (B) by positioning the wall near the top of the hillside. Use the soil removed at the base of the hill to fill in near the top of the wall.

Building Retaining Walls ▸

Backfill with crushed stone and install a perforated drain pipe about 6" above the bottom of the backfill. Vent the pipe to the side or bottom of the retaining wall, where runoff water can flow away from the hillside without causing erosion.

Make a stepped trench when the ends of a retaining wall must blend into an existing hillside. Retaining walls often are designed so the ends curve or turn back into the slope.

How to Build a Retaining Wall Using Interlocking Block

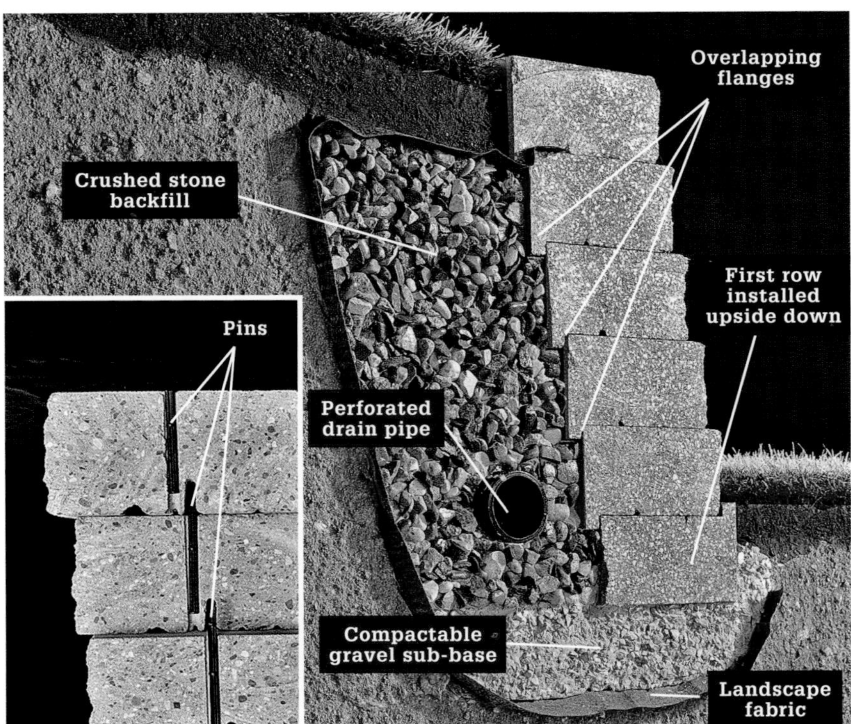

Interlocking wall blocks do not need mortar. Some types are held together with a system of overlapping flanges that automatically set the backward pitch (batter) as the blocks are stacked, as shown in this project. Other types of blocks use fiberglass pins (inset).

Excavate the hillside, if necessary. Allow 12" of space for crushed stone backfill between the back of the wall and the hillside. Use stakes to mark the front edge of the wall. Connect the stakes with mason's string, and use a line level to check for level.

Dig out the bottom of the excavation below ground level, so it is 6" lower than the height of the block. For example, if you use 6"-thick block, dig down 12". Measure down from the string to make sure the bottom base is level.

Line the excavation with strips of landscape fabric cut 3 ft. longer than the planned height of the wall. Make sure all seams overlap by at least 6".

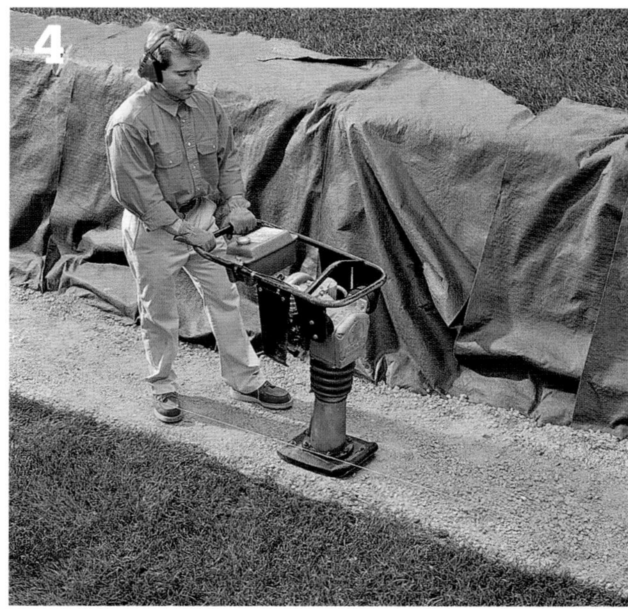

Spread a 6" layer of compactable gravel over the bottom of the excavation as a sub-base and pack it thoroughly. A rented tamping machine, or jumping jack, works better than a hand tamper for packing the sub-base.

Lay the first course of block, aligning the front edges with the mason's string. (When using flanged block, place the first course upside down and backward.) Check frequently with a level, and adjust, if necessary, by adding or removing sub-base material below the blocks.

Lay the second course of block according to manufacturer's instructions, checking to make sure the blocks are level. (Lay flanged block with the flanges tight against the underlying course.) Add 3 to 4" of gravel behind the block, and pack it with a hand tamper.

Make half-blocks for the corners and ends of a wall and use them to stagger vertical joints between courses. Score full blocks with a circular saw and masonry blade, and then break the blocks along the scored line with a maul and chisel.

(continued)

Add and tamp crushed stone, as needed, to create a slight downward pitch (about ¼" of height per foot of pipe) leading to the drain pipe outlet. Place the drain pipe on the crushed stone, 6" behind the wall, with the perforations face down. Make sure the pipe outlet is unobstructed. Lay courses of block until the wall is about 18" above ground level, staggering the vertical joints.

Fill behind the wall with crushed stone, and pack it thoroughly with the hand tamper. Lay the remaining courses of block, except for the cap row, backfilling with crushed stone and packing with the tamper as you go.

Before laying the cap block, fold the end of the landscape fabric over the crushed stone backfill. Add a thin layer of topsoil over the fabric, and then pack it thoroughly with a hand tamper. Fold any excess landscape fabric back over the tamped soil.

Apply construction adhesive to the top course of block, and then lay the cap block. Use topsoil to fill in behind the wall and to fill in the base at the front of the wall. Install sod or plants as desired.

How to Add a Curve to an Interlocking Block Retaining Wall

Outline the curve by first driving a stake at each end and then driving another stake at the point where lines extended from the first stakes would form a right angle. Tie a mason's string to the right-angle stake, extended to match the distance to the other two stakes, establishing the radius of the curve. Mark the curve by swinging flour or spray paint at the string end, like a compass.

Excavate for the wall section, following the curved layout line. To install the first course of landscape blocks, turn them upside down and backwards and align them with the radius curve. Use a 4-ft. level to ensure the blocks sit level and are properly placed.

Install subsequent courses so the overlapping flange sits flush against the back of the blocks in the course below. As you install each course, the radius will change because of the backwards pitch of the wall, affecting the layout of the courses. Where necessary, trim blocks to size. Install using landscape construction adhesive, taking care to maintain the running bond.

Use half blocks or cut blocks to create finished ends on open ends of the wall.

Outdoor Kitchen

With its perfect blend of indoor convenience and alfresco atmosphere, it's easy to see why the outdoor kitchen is one of today's most popular home upgrades. In terms of design, outdoor kitchens can take almost any form, but most are planned around the essential elements of a built-in grill and convenient countertop surfaces (preferably on both sides of the grill). Secure storage inside the cooking cabinet is another feature many outdoor cooks find indispensable.

The kitchen design in this project combines all three of these elements in a moderately sized cooking station that can fit a variety of kitchen configurations. The structure is freestanding and self-supporting, so it can go almost anywhere—on top of a patio, right next to a house wall, out by the pool, or out in the yard to create a remote entertainment getaway. Adding a table and chairs or a casual sitting area might be all you need to complete your kitchen accommodations. But best of all, this kitchen is made almost entirely of inexpensive masonry materials.

Concrete and masonry are ideally suited to outdoor kitchen construction. Both are noncombustible, not damaged by water, and can easily withstand decades of outdoor exposure. In fact, a little weathering makes masonry look even better. In this project, the kitchen's structural cabinet is built with concrete block on top of a reinforced concrete slab. The countertop is two-inch-thick poured concrete that you cast in place over two layers of cementboard.

Tools & Materials ▸

- Chalk line
- Pointing trowel
- Masonry mixing tools
- Level
- Mason's string
- Circular saw with masonry blade
- Eye protection and work gloves
- Utility knife
- Straightedge
- Square-notched trowel
- Metal snips
- Wood float
- Steel trowel
- Drill with masonry bit
- Mortar mix or mason mix
- Concrete block
- Reinforcing wire or rebar
- Hammer
- Metal reinforcement
- Steel angle iron
- ½" cementboard
- Lumber (2 × 4, 2 × 6)
- Deck screws (2½, 3")
- Silicone caulk
- Stucco lath
- Vegetable oil or other release agent
- Concrete mix
- Base coat stucco
- Finish coat stucco
- Sealer
- Sandpaper

This practical outdoor kitchen has just what the serious griller needs—a built-in grill and plenty of countertop space for preparing and serving meals. At just over 8 ft. long and about 3 ft. wide, the kitchen can fit almost anywhere on a standard concrete patio.

Construction Details

The basic structure of this kitchen consists of five courses of standard 8 × 8 × 16" concrete block. Two mortared layers of ½" cementboard serve as a base for the countertop. The 2"-thick poured concrete layer of the countertop extends 1½" beyond the rough block walls and covers the cementboard edges. The walls receive a two-coat stucco finish that can be tinted during the mixing or painted after it cures. Doors in the front of the cabinet provide access to storage space inside and to any utility connections for the grill. The kitchen's dimensions can easily be adjusted to accommodate a specific location, cooking equipment, or doors and additional amenities.

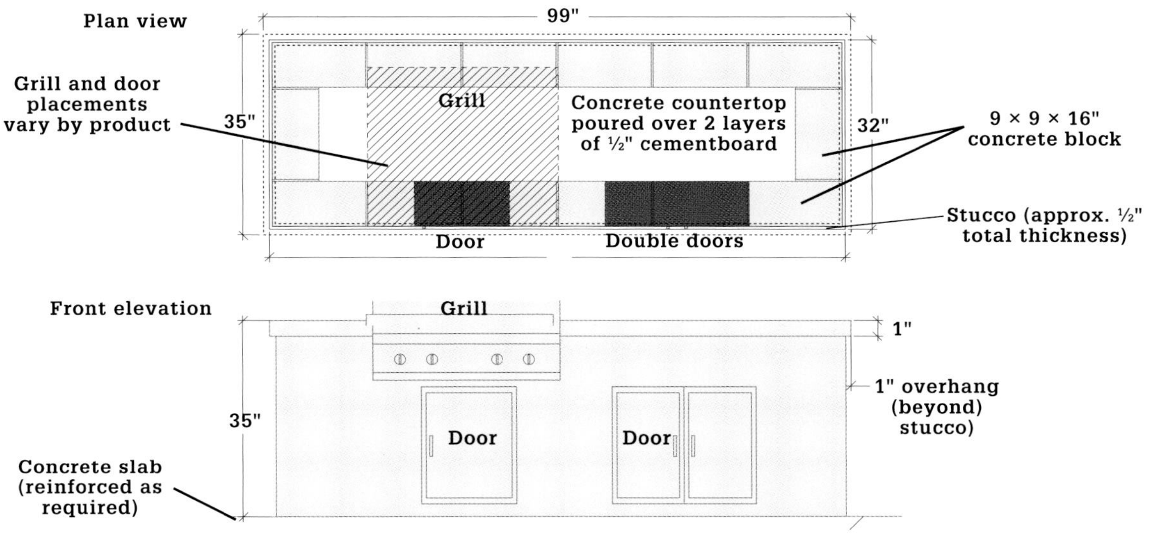

Planning an Outdoor Kitchen Project ▸

Whether you model your project after the one shown here or create your own design, there are a few critical factors to address as part of your initial planning:

Foundation: Check with your local building department about foundation requirements for your kitchen. Depending on the kitchen's size and location, you may be allowed to build on top of a standard four-inch-thick reinforced concrete patio slab, or you might need frost footings or a reinforced "floating footing" similar to the one shown on page 148 (Brick Barbecue).

Grill and door units: You'll need the exact dimensions of the grill, doors, and any other built-in features before you draw up your plans and start building. When shopping for equipment, keep in mind its utility requirements and the type of support system needed for the grill and other large units. Some grills are drop-in and are supported only by the countertop; others must be supported below with a noncombustible, load-bearing material such as concrete block or a poured concrete platform.

A grill gas line typically extends up into the cabinet space under the grill and is fitted with a shutoff valve.

Utility hookups: Grills fueled by natural gas require a plumbed gas line, and those with electric starters need an outdoor electrical circuit, both running into the kitchen cabinet. To include a kitchen sink, you'll need a dedicated water line and a drain connection (to the house system, directly to the city sewer, or possibly to a dry well on your property). Outdoor utilities are strictly governed by building codes, so check with the building department for requirements. Generally, the rough-in work for utilities is best left to professionals.

How to Build an Outdoor Kitchen

Pour the foundation or prepare the slab for the wall construction. See pages 46 to 49 for help with building frost footings and pages 68 to 71 for pouring a concrete slab. To prepare an existing slab, clean the surface thoroughly to remove all dirt, oils, concrete sealers, and paint that could prevent a good bond with mortar.

Dry-lay the first course of block on the foundation to test the layout. If desired, use 2- or 4"-thick solid blocks under the door openings. Snap chalk lines to guide the block installation, and mark the exact locations of the door openings.

Set the first course of block into mortar, following the basic techniques shown on pages 124 to 127. Cut blocks as needed for the door openings. Lay the second course, offsetting the joints with the first course in a running-bond pattern.

Continue laying up the wall, adding reinforcing wire or rebar if required by local building code. Instead of tooling the mortar joints for a concave profile, use a trowel to slice excess mortar from the blocks. This creates a flat surface that's easier to cover with stucco.

Install steel angle lintels to span over the door openings. If an opening is in line with a course of block, mortar the lintels in place on top of the block. Otherwise, use a circular saw with a masonry blade to cut channels for the horizontal leg of the angle. Lintels should span 6" beyond each side of an opening. Slip the lintel into the channels, and then fill the block cells containing the lintel with mortar to secure the lintel in place. Lay a bed of mortar on top of the lintels, and then set block into the mortar. Complete the final course of block in the cabinet and let the mortar cure.

Cut two 8-ft.-long sheets of cementboard to match the outer dimensions of the block cabinet. Apply mortar to the tops of the cabinet blocks and then set one layer of cementboard into the mortar. If you will be installing a built-in grill or other accessories, make cutouts in the cementboard with a utility knife or a jigsaw with a remodeler's blade.

Cut pieces to fit for a second layer of cementboard. Apply a bed of mortar to the top of the first panel and then lay the second layer pieces on top, pressing them into the mortar so the surfaces are level. Let the mortar cure.

(continued)

To create a 1½" overhang for the countertop, build a perimeter band of 2 × 4 lumber; this will serve as the base of the concrete form. Cut the pieces to fit tightly around the cabinet along the top. Fasten the pieces together at their ends with 3" screws so their top edges are flush with the bottom of the cementboard.

Cut vertical 2 × 4 supports to fit snugly between the foundation and the bottom of the 2 × 4 band. Install a support at the ends of each wall and evenly spaced in between. Secure each support with angled screws driven into the band boards.

Build the sides of the countertop form with 2 × 6s cut to fit around the 2 × 4 band. Position the 2 × 6s so their top edges are 2" above the cementboard and fasten them to the band with 2½" screws.

Form the opening for the grill using 2 × 6 side pieces (no overhang inside opening). Support the edges of the cementboard along the grill cutout with cleats attached to the 2 × 6s. Add vertical supports as needed under the cutout to keep the form from shifting under the weight of the concrete.

Cut a sheet of stucco lath to fit into the countertop form, leaving a 2" space along the inside perimeter of the form. Remove the lath and set it aside. Seal the form joints with a fine bead of silicone caulk and smooth with a finger. After the caulk dries, coat the form boards (not the cementboard) with vegetable oil or other release agent.

Dampen the cementboard with a mist of water. Mix a batch of countertop mix, adding color if desired (see page 104). Working quickly, fill along the edges of the form with concrete, carefully packing it down into the overhang portion by hand.

Fill the rest of the form halfway up with an even layer of concrete. Lay the stucco lath on top, and then press it lightly into the concrete with a float. Add the remaining concrete so it's flush with the tops of the 2 × 6s.

Tap along the outsides of the form with a hammer to remove air bubbles trapped against the inside edges. Screed the top of the concrete with a straight 2 × 4 riding along the form sides. Add concrete as needed to fill in low spots so the surface is perfectly flat.

(continued)

After the bleed water disappears, float the concrete with a wood or magnesium float. The floated surface should be flat and smooth but will still have a somewhat rough texture. Be careful not to overfloat and draw water to the surface.

A few hours after floating, finish the countertop as desired. A few passes with a steel finishing trowel yields the smoothest surface. Hold the leading edge of the trowel up and work in circular strokes. Let the concrete set for a while between passes.

Moist-cure the countertop with a fine water mist for three to five days. Remove the form boards. If desired, smooth the countertop edges with an abrasive brick and/or a diamond pad or sandpaper. After the concrete cures, apply a food-safe sealer to help prevent staining.

Prepare for door installation in the cabinet. Outdoor cabinet doors are usually made of stainless steel and typically are installed by hanging hinges or flanges with masonry anchors. Drill holes for masonry anchors in the concrete block, following the door manufacturer's instructions.

Quick Tip ▸

Honeycombs or air voids can be filled using a cement slurry of cement and water applied with a rubber float. If liquid cement color was used in your countertop concrete mix, color should be added to the wet cement paste. Some experimentation will be necessary.

Finish installing and hanging the doors. Test the door operations and make sure to caulk around the edges with high-quality silicone caulk. *Note: Doors shown here are best installed before the stucco finish is applied to the cabinet. Other doors may be easier to install following a different sequence.*

To finish the cabinet walls, begin by dampening the contrete block and then applying a ⅜"-thick base coat of stucco, following the steps on pages 250 to 255. Apply an even layer over the walls; then smooth the surface with a wood float and moist-cure the stucco for 48 hours or as directed by the manufacturer.

Apply a finish coat of tinted stucco that's at least ⅛" thick. Evenly saturate the base coat stucco surface with water prior to applying the the finish coat. Texture the surface as desired. Moist-cure the stucco for several days as directed.

Set the grill into place, make the gas connection, and then check it carefully for leaks. Permanently install the grill following the manufacturer's directions. The joints around grills are highly susceptible to water intrusion; seal them thoroughly with an approved caulk to help keep moisture out of the cabinet space below.

Brick Barbecue

The barbecue design shown here is constructed with double walls—an inner wall, made of heat-resistant fire brick set on edge, surrounding the cooking area, and an outer wall, made of engineer brick. We chose this brick because its larger dimensions mean you'll have fewer bricks to lay. You'll need to adjust the design if you select another brick size. A four-inch air space between the walls helps insulate the cooking area. The walls are capped with thin pieces of cut stone.

Refractory mortar is recommended for use in areas of direct fire contact. It is heat resistant and the joints will last a long time without cracking. Ask a local brick yard to recommend a refractory mortar for outdoor use.

Mortar Data ▸

Type N Mortar: Nonstructural mortar for veneer applications, reaches 750 psi @ 28 days

Type S Mortar: Structural mortar for veneer structural applications, exceeds 1,800 psi @ 28 days

The foundation combines a 12-inch-deep footing supporting a reinforced slab. This structure, known as a floating footing, is designed to shift as a unit when temperature changes cause the ground to shift. Ask a building inspector about local building code specifications.

Tools & Materials ▸

Tape measure
Hammer
Brickset chisel
Mason's string
Shovel
Aviation snips
Reciprocating saw or hacksaw
Masonry hoe
Shovel
Wood float
Chalk line
Level
Wheelbarrow
Mason's trowel
Jointing tool
Garden stakes
2 × 4 lumber
18-gauge galvanized metal mesh
#4 rebar
16-gauge tie wire
Bolsters
Fire brick
Engineer brick
Type N or Type S mortar
⅜"-dia. dowel
Metal ties
4" tee plates
Brick sealer
Stainless-steel expanded mesh
Cooking grills
Ash pan
Concrete mix
Work gloves

The addition of a brick barbecue to your patio is a non-intrusive way to incorporate summer cooking into a four-seasons space, while adding the beauty and stability of brick to your outdoor decorating scheme.

Pouring a Floating Footing ▸

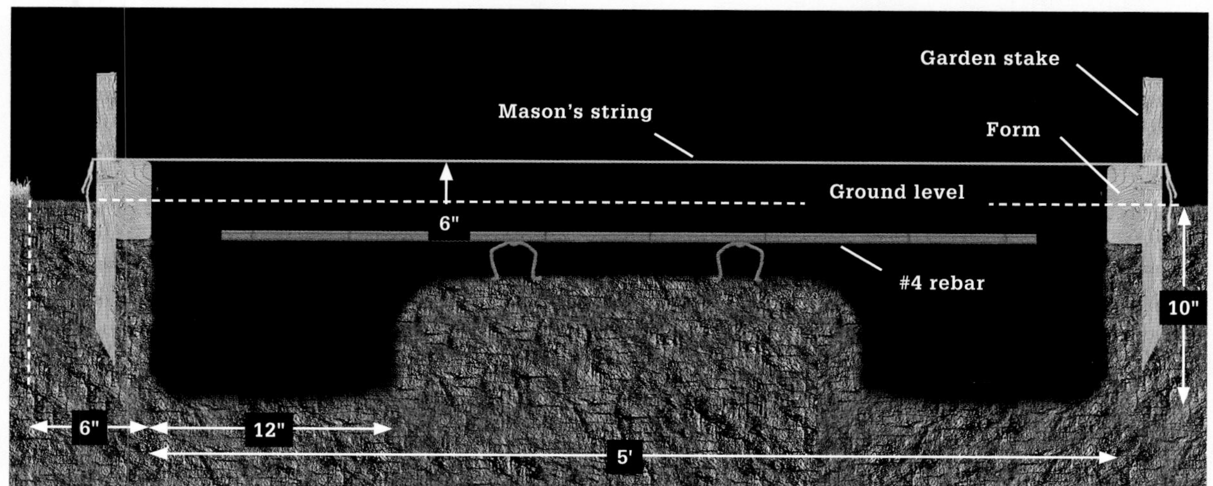

Lay out a 4 × 5-ft. area. Dig a continuous trench 12" wide × 10" deep along the perimeter of the area, leaving a rectangular mound in the center. Remove 4" of soil from the top of the mound, and round over the edges. Set a 2 × 4 form around the site so that the top is 2" above the ground along the back and 1½" above the ground along the front. This slope will help shed water. Reinforce the footing with five 52"-long pieces of rebar. Use a mason's string and a line level to ensure that the forms are level from side to side. Set the rebar on the bolster 4" from the front and rear sides of the trench, centered from side to side. Space the remaining three bars evenly in between. Coat the forms with vegetable oil or release agent and pour the concrete.

How to Build a Brick Barbecue

After the footing has cured for one week, use a chalk line to mark the layout for the inner edge of the fire brick wall. Make a line 4" in from the front edge of the footing and a center line perpendicular to the first line. Make a 24 × 32" rectangle that starts at the 4" line and is centered on the center line.

Dry-lay the first course of fire brick around the outside of the rectangle, allowing for ⅛"-thick mortar joints. *Note: Proper placement of the inner walls is necessary so they can support the grills. Start with a full brick at the 4" line to start the right and left walls. Complete the course with a cut brick in the middle of the short wall.*

(continued)

Dry-lay the outer wall, as shown here, using 4 × 3⅕ × 8" nominal engineer brick. Gap the bricks for ⅜" mortar joints. The rear wall should come within ⅜" of the last fire brick in the left inner wall. Complete the left wall with a cut brick in the middle of the wall. Mark reference lines for this outer wall.

Make a story pole. On one side, mark 8 courses of fire brick, leaving a ⅜" gap for the bottom mortar joint and ⅛" gaps for the remaining joints. The top of the final course should be 36" from the bottom edge. Transfer the top line to the other side of the pole. Lay out 11 courses of engineer brick, spacing them evenly so that the final course is flush with the 36" line. Each horizontal mortar joint will be slightly less than ½" thick.

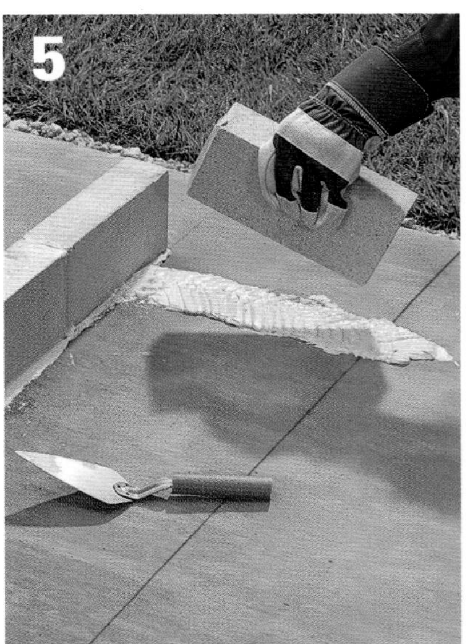

Lay a bed of mortar for a ⅜" joint along the reference lines for the inner wall, and then lay the first course of fire brick using ⅛" joints between the bricks.

Lay the first course of the outer wall, using Type N or Type S mortar. Use oiled ⅜" dowels to create weep holes behind the front bricks of the left and right walls. Alternate laying the inner and outer walls, checking your work with the story pole and a level after every course.

Start the second course of the outer wall using a half-brick butted against each side of the inner wall, and then complete the course. Because there is a half-brick in the right outer wall, you need to use two three-quarter bricks in the second course to stagger the joints.

Place metal ties between the corners of the inner and outer walls at the second, third, fifth, and seventh courses. Use ties at the front junctions and along the rear walls. Mortar the joint where the left inner wall meets the rear outer wall.

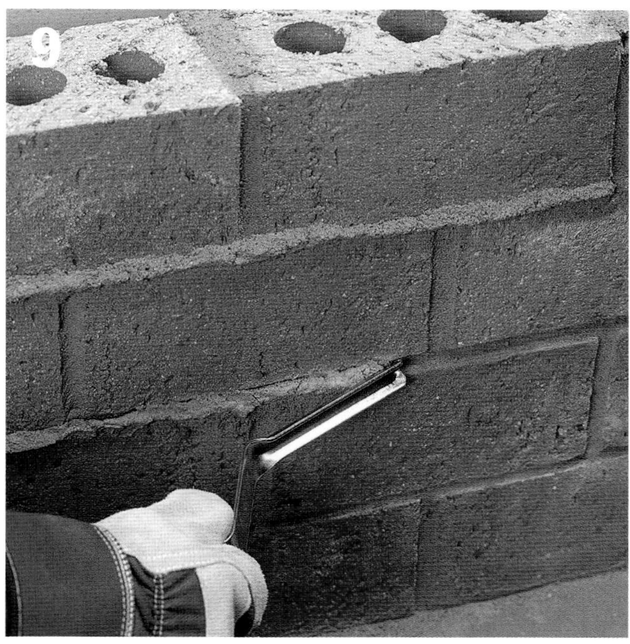

Smooth the mortar joints with a jointing tool when the mortar has hardened enough to resist minimal finger pressure. Check the joints in both walls after every few courses. The different mortars may need smoothing at different times.

Add tee plates for grill supports above the fifth, sixth, and seventh courses. Use 4"-wide plates with flanges that are no more than 3⁄32" thick. Position the plates along the side fire brick walls, centered 3", 12", 18", and 27" from the rear fire brick wall.

When both walls are complete, install the capstones. Lay a bed of Type N or Type S mortar for a 3⁄8"-thick joint on top of the inner and outer walls. Lay the capstone flat across the walls, keeping one end flush with the inner face of the fire brick. Make sure the bricks are level and tool the joints when they are ready. After a week, seal the capstones and the joints between them with brick sealer and install the grills.

Brick Planter

Brick is the masonry material of choice for elegant entry planters. It also complements a brick-paver landing. For a foundation for the planter, pour a slab that is separated from adjacent structures, such as a landing or house foundation, by isolation joints. With larger planter projects, a frost footing often is required; check your local building code.

Your planter can capture the rustic beauty of an antique trough or dress up a brick entryway.

Tools & Materials ▸

- Mason's string
- Line level
- Drill
- Level
- Shovel
- Rake
- Hoe
- Wheelbarrow
- Hand tamper
- Rubber mallet
- Jointing tool
- Tape measure
- Broom
- Mason's trowel
- Type S mortar
- Bricks
- Screws
- 1 × 4 lumber
- Stakes
- 1 × 4 concrete step forms
- Pavers
- Compactable gravel
- 3/8"-dia. copper or PVC tubing
- Sand
- Isolation board
- Cap bricks
- Landscape fabric
- Work gloves

How to Build a Brick Planter

Excavate the building site, install forms and isolation boards, and pour a concrete base for the project. Let the foundation cure for three days before building on it. Remove forms, and then trim isolation boards so they are level with the tops of adjoining structures, like the landing shown above. *Tip: Cover adjoining surfaces for protection.*

Test-fit the first course of the project, and then outline the project on the concrete surface. Dampen the surface slightly then mix mortar and throw a mortar bed in one corner (pages 118 to 121). Begin laying bricks for the project, buttering the exposed end of each brick before setting it.

Lay one section of the first course, checking the bricks frequently with a level to make sure the tops are level and even. Lay two corner return bricks perpendicular to the end bricks in the first section and use a level to make sure they are even across the tops.

Install weep holes for drainage in the first course of bricks on the sides farthest away from permanent structures. Cut ⅜"-dia. copper or PVC tubing about ¼" longer than the width of one brick and set the pieces into the mortar joints between bricks, pressing them into the mortar bed so they touch the footing. Make sure mortar doesn't block the openings.

Finish building all sides of the first course. Lay the second course of bricks, reversing the directions of the corner bricks to create staggered vertical joints if using a running-bond pattern. Fill in brick courses to full height, building up one course at a time. Check frequently to make sure the tops of the bricks are level and sides are plumb.

Install cap bricks to keep water from entering the cores of the brick and to enhance the visual effect. Set the cap bricks into a ⅜"-thick mortar bed, buttering one end of each cap brick. Let the mortar cure for one week. Before adding soil, pour a 4 to 6"-thick layer of gravel into the bottom of the planter for drainage, and then line the bottom and sides of the planter with landscape fabric to prevent dirt from running into and clogging the drainage tubes.

Brick Pillars

Decorative pillars are easy to design because you don't have to be concerned about the seasonal shifting of attached brick or stone walls. We designed a pair of 12 × 16" pillars using only whole bricks, so you don't need to worry about splitting or cutting. These pillars are refined in appearance but sturdy enough to last for decades.

Once the last course of bricks is in place, you can add a brick or stone cap for a finished look. Or, build two pillars connected by an arch (pages 156 to 159). If you're planning an arch, consider attaching hardware for an iron gate. It's far easier to place the hardware in fresh mortar, so make a note of the brick courses where the hardware will go. The settings will look cleaner this way and the hardware will stay secure for a long time.

Tools & Materials ▸

- Level
- Bricklayer's trowel
- Jointing tool
- Aviation snips
- Wheelbarrow
- Shovel
- Hoe
- Tape measure
- Standard modular bricks (4 × 2⅔ × 8")
- Pointing chisel
- Dowel
- Type N mortar mix
- ¼" wire mesh
- Capstone or concrete cap
- Lumber (1 × 3, 2 × 2)
- ⅜"-thick wood scraps
- Work gloves

This 4-ft. pillar was built with 18 courses of brick. A brick cap adds a touch of elegance and protects against rain, ice, and snow. You can also build pillars with stone caps, or, as shown on the following pages, use cast concrete caps, which are available in many sizes.

Pour footings that are 4" longer and wider than the pillars on each side. This project calls for 16 × 20" footings.

Building Brick Pillars ▸

Use a story pole to maintain consistent mortar joint thickness. Line up a scrap 1 × 2 on a flat tabletop alongside a column of bricks, spaced ⅜" apart. Mark the identical spacing on the 1 × 2. Hold up pole after every few courses to check the mortar joints for consistent thickness.

Cut a straight 2 × 2 to fit tight in the space between the two pillars. As you lay each course for the second pillar, use the 2 × 2 to check the span.

How to Build Brick Pillars

Once the footing has cured, dry-lay the first course of five bricks, centered on the footing. Mark reference lines around the bricks.

Lay a bed of mortar inside the reference lines and lay the first course.

(continued)

Use a pencil or dowel coated with vegetable oil to create a weep hole in the mortar in the first course of bricks. The hole ensures drainage of any moisture that seeps into the pillar.

Lay the second course, rotating the pattern 180°. Lay additional courses, rotating the pattern 180° with each course. Use the story pole and a level to check each face of the pillar after every other course. It's important to check frequently, since any errors will be exaggerated with each successive course.

After every fourth course, cut a strip of ¼" wire mesh and place it over a thin bed of mortar. Add another thin bed of mortar on top of the mesh, and then add the next course of brick.

After every five courses, use a jointing tool to smooth the joints that have hardened enough to resist minimal finger pressure.

For the final course, lay the bricks over a bed of mortar and wire mesh. After placing the first two bricks, add an extra brick in the center of the course. Lay the remainder of the bricks around it. Fill the remaining joints, and work them with the jointing tool as soon as they become firm.

Build the second pillar in the same way as the first. Use the story pole and measuring rod to maintain identical dimensions and spacing.

How to Install a Cap Stone

Select a capstone 3" longer and wider than the top of the pillar. Mark reference lines on the bottom for centering the cap. Do not install caps if you are adding an arch to your pillars.

Spread a ½"-thick bed of mortar on top of the pillar. Center the cap on the pillar using the reference lines. Strike the mortar joint under the cap so it's flush with the pillar. *Note: If mortar squeezes out of the joint, press ⅜"-thick wood scraps into the mortar at each corner to support the cap. Remove the scraps after 24 hours and fill in the gaps with mortar.*

Brick Archway

Building an arch over a pair of pillars is a challenging task made easier with a simple, semi-circular plywood form. With the form in place, you can create a symmetrical arch by laying bricks along the form's curved edge. Select bricks equal in length to those used in the pillars.

Brick or concrete archways were very common during the Roman Empire, when clever engineers devised ways to leverage geometry in their favor while constructing baths, aqueducts, and many other masonry structures. The principles are so sound and the materials so durable that many ancient archways remain standing today.

Tools & Materials ▸

- Hammer
- Joint chisel
- Mason's hammer
- Pry bar
- Jigsaw
- Circular saw
- Drill
- Compass
- Level
- Mason's string
- Trowel
- Jointing tool
- Tuck-pointer
- ¾" plywood
- ¼" plywood
- Wallboard screws (1 and 2")
- Bricks
- Type N mortar mix
- 2 × 4 and 2 × 8 lumber
- Shims
- Work gloves

If you're building an arch over existing pillars, measure the distance between the pillars at several points. The span must be the same at each point in order for the pillars to serve as strong supports for your arch.

Build a Form for an Arch ▸

1. Determine the distance between the inside edges of the tops of your pillars. Divide the distance in half and then subtract ¼". Use this as the radius in Step 2.
2. Mark a point at the center of a sheet of ¾" plywood. Use a pencil and a piece of string to scribe a circle on the plywood using the radius calculated in Step 1. Cut out the circle with a jigsaw. Mark a line through the center point of the circle and cut the circle in half.
3. Construct the form by bracing the two semicircles using 2" wallboard screws and 2 × 4s. To calculate the length of the 2 × 4 braces, subtract the combined thickness of the plywood sheets —1½"—from the width of the pillars and cut the braces to length. Cover the top of the form with ¼" plywood, attached with 1" wallboard screws.

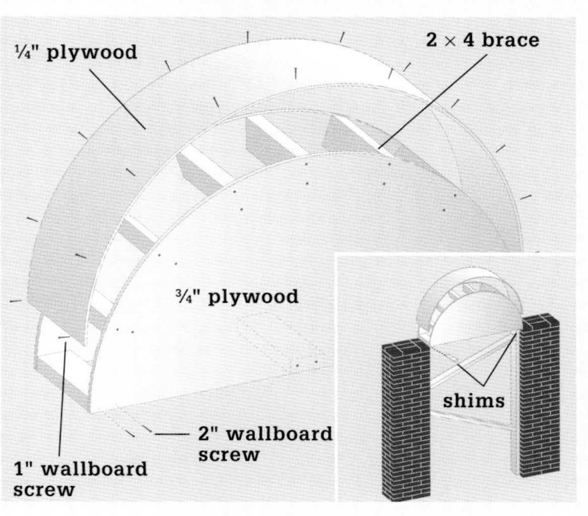

How to Build a Brick Arch

Tip: *If your pillars are capped, remove the caps before building an arch. Chip out the old mortar from underneath using a hammer and joint chisel. With a helper nearby to support the cap, use a pry bar and shims to remove each cap from the pillar.*

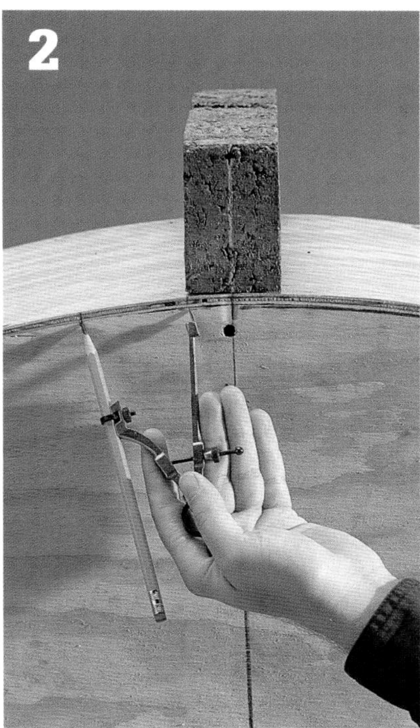

To determine brick spacing, start by centering a brick at the peak and placing a compass point at one edge. With the compass set to the width of one brick plus ¼", mark the form with the pencil.

Place the compass point on this new mark and make another mark along the curve. Continue making marks along the curve until less than a brick's width remains.

(continued)

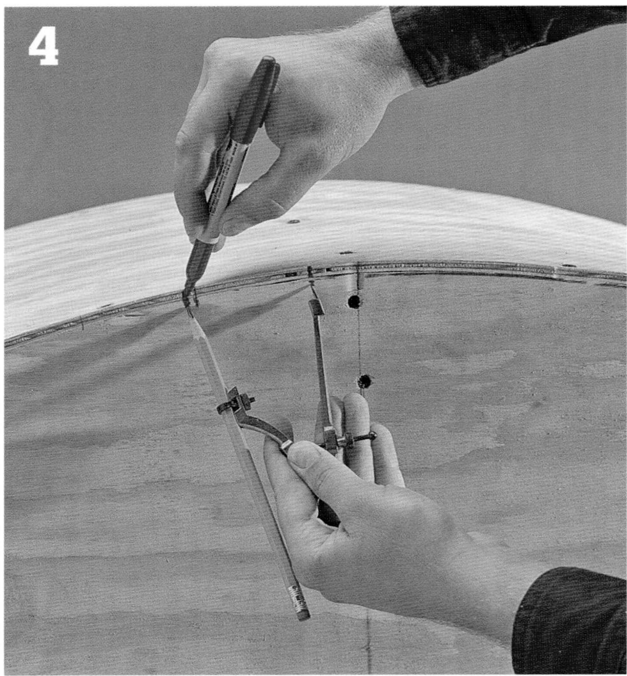

Divide this remaining width by the number of compass marks and increase the compass setting by this amount. Using a different color, make final reference marks to either side of the peak. Extend the pencil lines across the curved surface of the form and onto the far edge.

Cut two 2 × 8 braces ½" shorter than pillar height and prop one against each pillar with 2 × 4 cross braces. Place shims on top of each 2 × 8 to raise the form so its bottom is even with the tops of the pillars. Rest the plywood form on the braces.

Mix mortar and trowel a narrow ⅜" layer on top of one pillar. Place one brick, and then rap the top with a trowel handle to settle it. Butter the bottom of each subsequent brick and place it in position.

Place five bricks, and then tack a string to the center point of the form on each side and use the strings to check each brick's alignment. Take care not to dislodge other bricks as you tap a brick into position.

To balance the weight on the form, switch to the other side. Continue alternating until space for one brick remains. Smooth previous joints with a jointing tool as they become firm.

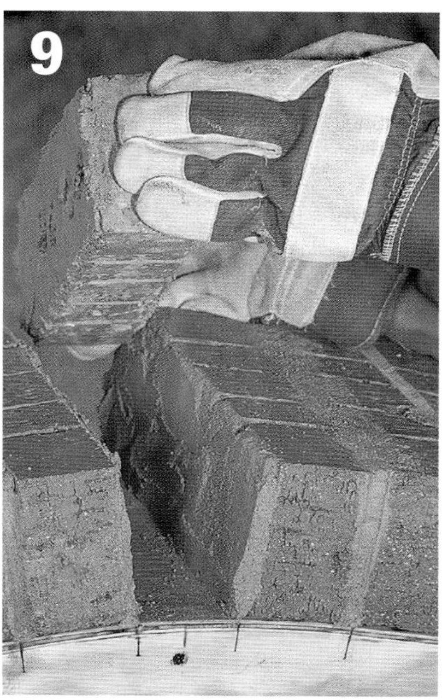

Butter the center, or keystone, brick as accurately as possible and ease it into place. Smooth the remaining joints with a jointing tool.

Lay a bed of mortar over the first course and then lay the second course halfway up each side, maintaining the same mortar joint thickness as in the first layer. Some of the joints will be staggered, adding strength to the arch.

Dry-lay several more bricks on one side—using shims as substitutes for mortar joints—to check the amount of space remaining. Remove the shims and lay the final bricks with mortar, and then smooth the joints with a jointing tool.

Leave the form in place for a week, misting occasionally. Carefully remove the braces and form. Tuck-point and smooth the joints on the underside of the arch.

Brick Wall Veneer

Brick veneer is essentially a brick wall built around the exterior walls of a house. It's attached to the house with metal wall ties and supported by a metal shelf hanger on the foundation. It's best to use queen-sized bricks for veneer projects because they're thinner than standard construction bricks. This means less weight for the house walls to support. Even so, brick veneer is quite heavy. Ask your local building inspector about building code rules that apply to your project. In the project shown here, brick veneer is installed over the foundation walls and side walls, up to the bottom of the windowsills on the first floor of the house. The siding materials in these areas are removed before installing the brick.

Construct a story pole before you start laying the brick so you can check your work as you go along to be sure your mortar joints are of a consistent thickness. A standard three-eighths-inch gap is used in the project shown here.

Tools & Materials ▸

- Pencil
- Hammer
- Circular saw
- Combination square
- Level
- Drill with masonry bit
- Socket wrench set
- Staple gun
- Mason's trowel
- Masonry hoe
- Mortar box
- Mason's chisel
- Maul
- Pressure-treated 2 × 4s
- ⅜ × 4" lag screws and washers
- 10d Nails
- Mason's string
- Silicone caulk
- 2 × 2
- Lead sleeve anchors
- Angle iron for metal shelf supports
- 30-mil PVC roll flashing
- Corrugated metal wall ties
- Brickmold for sill extensions
- Sill-nosing trim
- Type N mortar
- Bricks
- ⅜"-dia. cotton rope
- Work gloves and eye protection

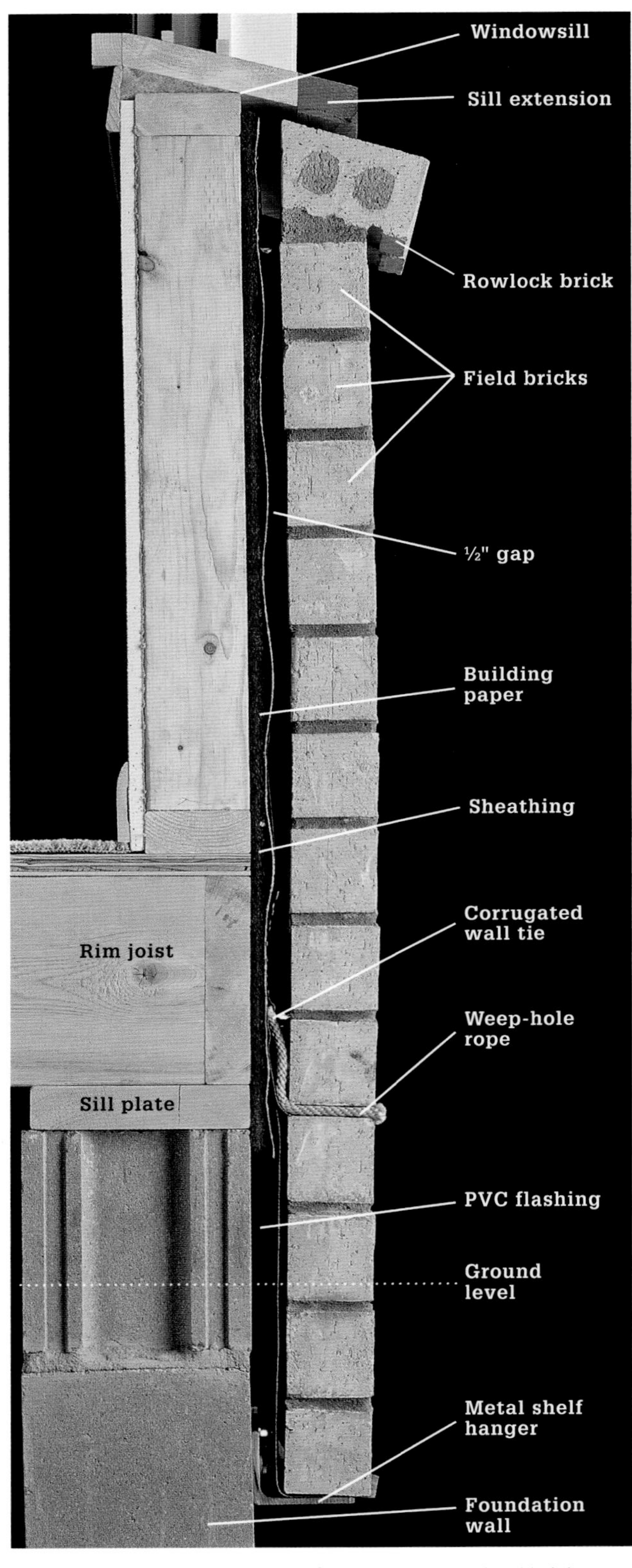

Anatomy of a brick veneer facade: Queen-sized bricks are stacked onto a metal or concrete shelf and connected to the foundation and walls with metal ties. Rowlock bricks are cut to follow the slope of the windowsills, and then laid on edge over the top course of bricks.

How to Install Brick Veneer

Remove all siding materials in the area you plan to finish with brick veneer. Before laying out the project, cut the sill extension from a pressure-treated 2 × 4. Tack the extension to the sill temporarily.

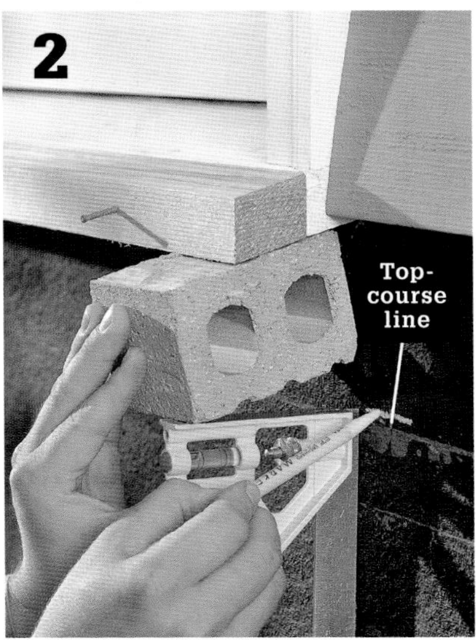

Precut the bricks to follow the slope of the sill and overhang the field brick by 2". Position this rowlock brick directly under the sill extension. Use a combination square or level to transfer the lowest point on the brick onto the sheathing (marking the height for the top course of brick in the field). Use a level to extend the line. Remove the sill extensions.

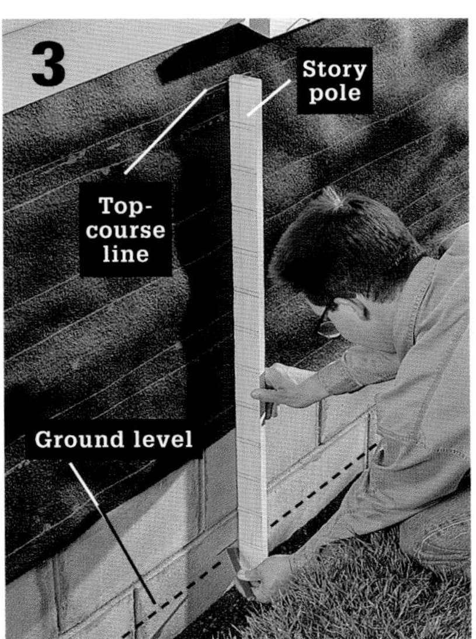

Make a story pole long enough to span the project area. Mark the pole with ⅜" joints between bricks. Dig a 12"-wide, 12"-deep trench next to the wall. Position the pole so the top-course line on the sheathing aligns with a top mark for a brick on the pole. Mark a line for the first course on the wall, below ground level.

Extend the mark for the first-course height across the foundation wall using a level as a guide. Measure the thickness of the metal shelf (usually ¼") and drill pilot holes for 10d nails into the foundation at 16" intervals along the first-course line, far enough below the line to allow for the thickness of the shelf. Slip nails into the pilot holes to create temporary support for the shelf.

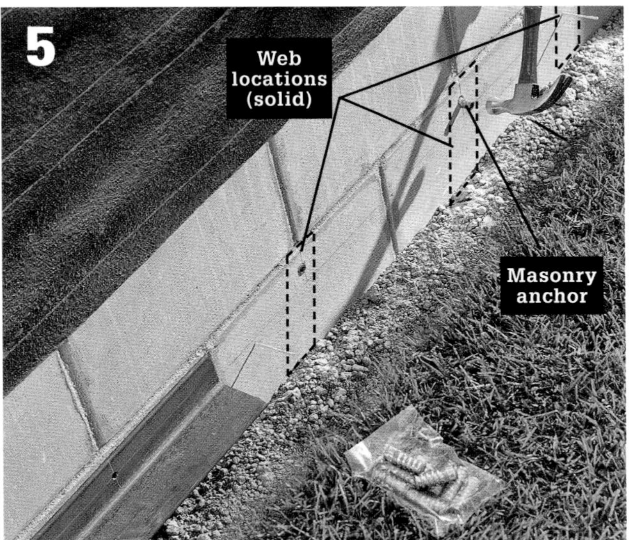

Set the metal shelf onto the temporary supports. Mark the location of the center web of each block onto the vertical face of the shelf. Remove the shelf and drill ⅜"-dia. holes for lag screws at the web marks. Set the shelf back onto the temporary supports and outline the predrilled holes on the blocks. Remove the shelf and drill holes for the masonry anchors into the foundation using a masonry bit. Drive masonry anchors into the holes.

(continued)

Reposition the shelf on the supports so the predrilled holes align with the masonry anchors. Attach the shelf to the foundation wall with ⅜ × 4" lag screws and washers. Allow ¹⁄₁₆" for an expansion joint between shelf sections. Remove the temporary support nails.

Staple 30-mil PVC flashing above the foundation wall after all sections of the metal shelf are attached. Be sure the PVC overlaps the metal shelf.

Test-fit the first course on the shelf. Work in from the ends, using spacers to set the gaps between bricks. You may need to cut the final brick for the course. Or, choose a pattern such as running bond that uses cut bricks.

Build up the corners two courses above ground level, and then attach line blocks and mason's string to the end bricks. Fill in the field bricks so they align with the strings. Every 30 minutes, smooth mortar joints that are firm.

Attach another course of PVC flashing to the wall so it covers the top course of bricks, and then staple building paper to the wall so it overlaps the top edge of the PVC flashing by at least 12". Mark wall-stud locations on the building paper.

Use the story pole to mark layout lines for the tops of every fifth course of bricks. Attach corrugated metal wall ties to the sheathing where the brick lines meet the marked wall-stud locations.

Fill in the next course of bricks, applying mortar directly onto the PVC flashing. At every third mortar joint in this course, tack a 10" piece of ⅜"-dia. cotton rope to the sheathing so it extends all the way through the bottom of the joint, creating a weep hole for drainage. Embed the metal wall ties in the mortar beds applied to this course.

Add courses of bricks, building up corners first, and then filling in the field. Embed the wall ties into the mortar beds as you reach them. Use corner blocks and a mason's string to verify the alignment, and check frequently with a 4-ft. level to make sure the veneer is plumb.

Apply a ½"-thick mortar bed to the top course and begin laying the rowlock bricks with the cut ends against the wall. Apply a layer of mortar to the bottom of each rowlock brick, and then press the brick up against the sheathing with the top edge following the slope of the windowsills.

Finish-nail the sill extensions to the windowsills. Nail sill-nosing trim to the siding to cover any gaps above the rowlock course. Fill cores of exposed rowlock blocks with mortar and caulk any gaps around the veneer with silicone caulk.

Mortared Brick Patio

Setting brick or concrete pavers into mortar is one of the most beautiful—and permanent—ways to dress up an old concrete slab patio. Mortared pavers arguably offer the most classic, finished appearance for either a backyard patio or a front entry stoop. The paving style used most often for mortared pavers is the standard running-bond pattern, which is also the easiest pattern to install.

Mortared pavers are appropriate for old concrete slabs that are flat, structurally sound, and relatively free of cracks. Minor surface flaws are generally acceptable, since the new mortar bed will compensate for slight imperfections. However, existing slabs with significant cracks or any evidence of shifting or other structural problems will most likely pass on those same flaws to the paver finish. For these, a nonmortared application is a safer solution. When in doubt, have your slab assessed by a qualified mason or concrete contractor to learn about your options. New concrete slabs are also suitable for mortared paving, but make sure the concrete has cured completely before applying the paver veneer.

Pavers for mortaring include natural clay brick units in both standard thickness (2⅝ inches) and thinner versions (1½ inches) and concrete pavers in various shapes and sizes. Any type you choose should be square-edged, to simplify the application and finishing of the mortar joints. When shopping for pavers, discuss your project with an expert masonry supplier. Areas that experience harsh winters call for the hardiest pavers available, graded SW or SX for severe weather. Also make sure the mortar you use is compatible with the pavers to minimize the risk of cracking and other problems.

Tools & Materials ▸

- Stiff brush or broom
- Circular saw
- Mason's trowel
- Mortar mixing tools
- 4-ft. level
- Rubber mallet
- Mortar bag
- Jointing tool
- Pointing trowel
- Concrete cleaner
- Pavers
- Spray bottle
- Isolation board
- ⅜ or ½" plywood
- Type S mortar
- Burlap
- Plastic sheeting
- Notched board
- Mason's string
- Straight 2 × 4
- Eye and ear protection
- Work gloves

Nothing dresses up an old concrete patio like brick or concrete pavers set in a bed of mortar. The mortaring process takes more time and effort than many finishing techniques, but the look is timeless and the surface is extremely durable.

How to Set Pavers in a Mortar Bed

Prepare the patio surface for mortar by thoroughly cleaning the concrete with a commercial concrete cleaner and/or a pressure washer. Make sure the surface is completely free of dirt, grease, oil, and waxy residue. ***Note: Follow manufacturer's instructions for proper use and safety.***

Dry-lay the border pavers along the edge of the patio slab. Gap the pavers to simulate the mortar joints using spacers cut from plywood equal to the joint thickness (⅜ or ½" is typical). Adjust the pavers as needed to create a pleasing layout with the fewest cuts possible. Install isolation board along the foundation wall if the paving abuts the house; this prevents the mortar from bonding with the foundation.

Mist the concrete with water to prevent premature drying of the mortar bed and then mix a batch of mortar as directed by the manufacturer.

Begin laying the border pavers by spreading a ½"-thick layer of mortar for three or four pavers along one edge of the patio using a mason's trowel. Lay the first few pavers, buttering the leading edge of each with enough mortar to create the desired joint thickness. Press or tap each paver in place to slightly compress the mortar bed.

Remove excess mortar from the tops and sides of the pavers. Use a level to make sure the pavers are even across the tops and check the mortar joints for uniform thickness. Tool the joints with a jointer as you go. Repeat the process to lay the remaining border pavers. Allow to dry as directed by the manufacturer.

Variation: To cover the edges of a raised slab, build wood forms similar to concrete forms (see pages 40 to 41). Set a gap between the forms and slab equal to the paver thickness plus ½". Install the edge pavers vertically or horizontally, as desired.

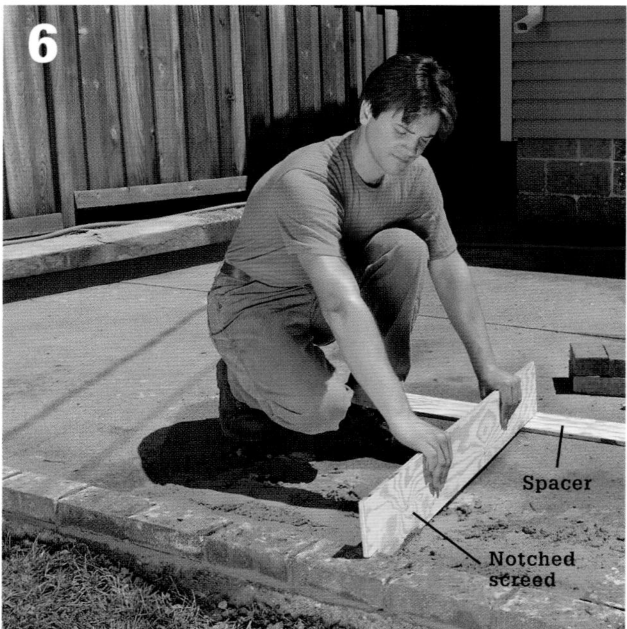

Spread and then screed mortar for the field pavers. Trowel on a ½"-thick layer of mortar inside the border, covering only about 3 or 4 sq. ft. to allow for working time before the mortar sets. Screed the mortar to a uniform ½" thickness using a notched board set atop the border pavers (set the interior end on a lumber spacer, as needed).

Begin laying the field pavers (without buttering them). Use the plywood spacers to set the gaps for mortar joints. Cut end pavers as needed. Keep the courses straight by setting the pavers along a string line referenced from the border pavers.

As you work, check the heights of the pavers with a level or a straight 2 × 4 to make sure all units are level with one another. If a paver is too high, press it down or tap it with a rubber mallet; if too low, lift it out, butter it's bottom with mortar, and reset it. Repeat steps 6 through 8 to complete the paver installation, and then let the mortar bed dry.

Fill the paver joints with fresh mortar using a mortar bag to keep the paver faces clean. Within each working section, fill the long joints between courses first, and then do the short joints between the paver ends. Overfill the joints slightly.

Tool the joints with a jointing tool—again, complete the long joints first, and then fill the next section. As the mortar begins to set (turns from glossy wet to flat gray) in each tooled section, scrape off excess mortar with a pointing trowel, being careful not to smear mortar onto the paver faces.

Let the mortar joints dry for a few hours (or as directed by the manufacturer) and then scrub the paver faces with a wet burlap rag to remove any excess mortar and residue. Cover the surface with plastic for 48 hours. Remove the plastic, and let the surface cure undisturbed for one week before using the patio.

Natural Stone

Stone is a natural landscaping material that is at home in any yard or garden. Elegant and graceful, natural stone weathers well and in many cases actually improves in appearance after it is installed. From tumbled textures with old-world appeal to slick, polished surfaces, stone can blend in quietly or make a bold design statement.

In this chapter:

Natural Stone Basics

The greatest challenges to working with natural stone are dealing with its weight and its lack of uniformity when it comes to size and shape. These challenges can be overcome with plenty of helpers, creativity, and determination. If you've never worked with natural stone, take home a few pieces and experiment before undertaking a large project.

Depending on the stone you choose, individual pieces may be quite heavy. The best time to recruit helpers and devise ramps and other lifting and towing devices is before you come across a stone that's too heavy to move into place. Stone supply yards frequently have specialty tools that you can rent or borrow. Those that don't offer that service undoubtedly will know local sources for such tools.

The unique nature of each piece makes it necessary to cut and sometimes shape or "dress" stones (pages 179). This often involves removing jagged edges and undesirable bumps, reshaping a stone to improve its stability within the structure, and cutting stones to fit.

Once the stones are cut and ready to put into position, working with them is much like working with any other placed masonry item.

How Much Do You Need? ▸

Before you visit a stoneyard, determine how much area you plan to cover—literally. Know the dimensions of your project: height, width, length. Convert this into square footage. For a wall, you will multiply the length and height to arrive at a measurement. This number will help a professional calculate how much stone you should buy.

Tucked into the crevices of this fieldstone sitting area are rock-loving plants like moss and creeping thyme that add color and interest to the space.

Working with Stone

A trip to a local stoneyard can be an inspiring start to any stonescaping project. While you're checking out the inventory, take notes, snap pictures if you must, and ask lots of questions of professionals. Keep in mind, when you source stone from outside your region, you will pay more for shipping and delivery. In most cases buying native rocks from local stoneyards is the way to go.

Aside from large stone pieces, gravel, sand, and black dirt are essential building materials. You can obtain these materials in poly bags at garden or building centers. For larger amounts, visit a landscape supply outlet and purchase them by the cubic yard or by the ton.

Gravel: This material is essentially a chip off the rock—mechanically broken-down stone that can be spread as surface material or used as a buffer in drainages zones, such as against a home. Gravel ranges in diameter from a quarter-inch to two inches. If gravel is river rock, its surface will be smooth, rounded, shiny, and variegated in color.

Sand: You'll generally find two types of sand: coarse building sand and fine sand for landscaping or sandboxes. Extremely fine sand used for sandblasting has no use in stonescaping or landscaping.

Black dirt: This material is usually sold in pulverized or sterilized forms. This is fine for backfilling, but if you are planting, you'll want to amend the soil. Dirt is normally sold by the cubic yard since the weight varies dramatically based on the moisture content.

Hauling Materials ▸

You can save delivery charges (usually $35 to $50) and control delivery times by hauling landscape materials yourself in a pickup or trailer. The yard workers at the supply center will load your vehicle free of charge with a front-end loader or skid loader. Do not overload your vehicle. Although most operators are aware of load limits, they will typically put in as much as you tell them to. As a general rule, a compact truck (roughly the size of a Ford Ranger) can handle one scoop of dirt, sand, or gravel, which is about three quarters of a cubic yard; a half-ton truck (Ford F-150) will take a scoop and a half (a little over a cubic yard), and a three-quarter ton truck (Ford-250) can haul two scoops (one and a half cubic yards) safely. Be sure to check the gross vehicle weight and payload data label on the driver's door.

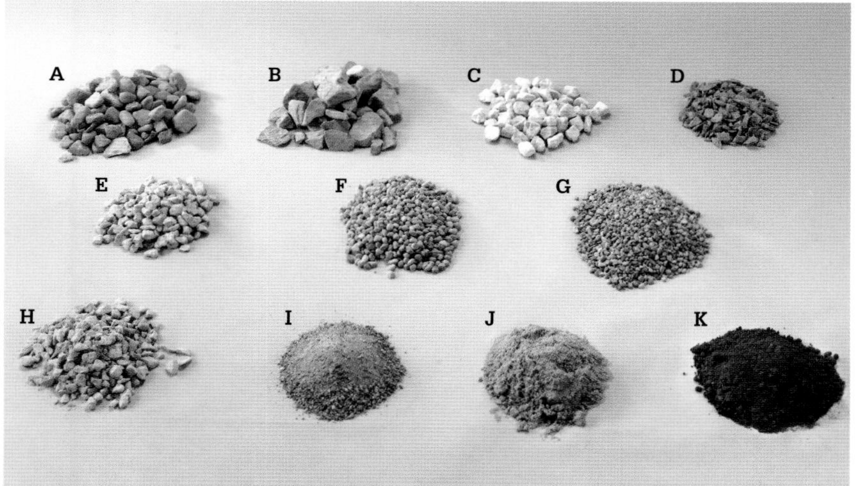

Loose materials used in stonescaping include: smooth river rock for top dressing (A); crushed rock for stable surface dressing (B); crushed quartz for decorative top dressing (C); trap rock for well-drained sub-bases or decorative top dressing (not a good choice for walking surfaces) (D); crushed limestone for sub-bases and top dressing (E); small-diameter river rock for top dressing and drainage (F); pea gravel for top dressing (G); compactable gravel (class V or class II) for compacted sub-bases (H); coarse sand for paver beds (I); fine sand for gap-filling (J); black dirt for backfilling (K).

Tools & Materials

Even though planet Earth naturally provides a wealth of rocks and dirt in various levels of solidity, natural stone can get fantastically expensive. So it pays to use a sharp pencil when you're estimating and sharp tools when you're working the stone to keep expensive waste to a minimum.

If you're doing the work yourself, don't expect to find six tons of premium ashlar at your local building center or gardening outlet. You'll need to do your shopping at a specialty landscape supplier. Where possible (and if you have access to an appropriate vehicle), select and haul the rock yourself to control quality and save delivery fees.

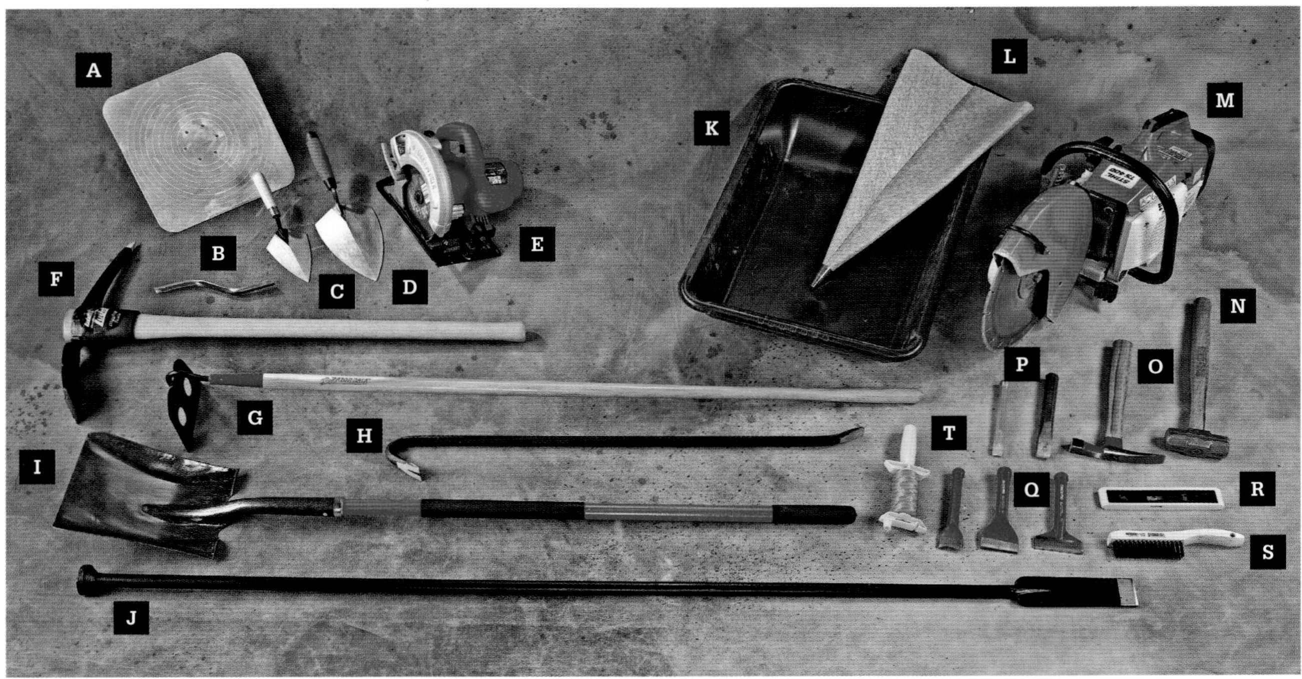

These tools are valuable helpers when working with natural stone. Mortar hawk (A); joining tool (B); pointing trowel (C); mason's trowel (D); circular saw with masonry blade (E); pick axe (F); masonry hoe (G); wrecking bar for prying small to medium stones (H); square-end spade (I); heavy-duty spud bar for prying larger stones (J); mortar box (K); mortar bag (L); masonry saw (M); hand maul (N); brick hammer (O); stone chisels (P); bricksets (Q); torpedo level (R); wire brush (S); mason's string (T).

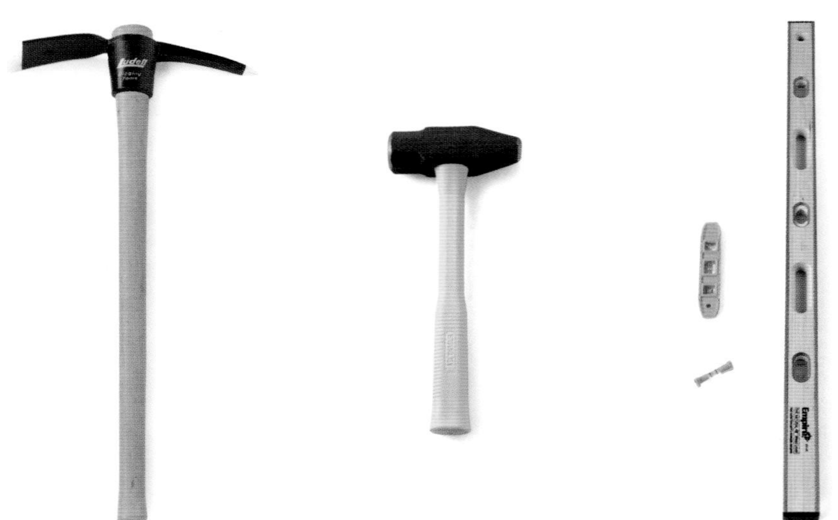

From left to right: a Mattock is similar to a pick axe and is used for digging or breaking up hard ground. Also called a grub hoe. A 3-lb. hand maul can be used to crush stone, dress ragged stone edges, or drive bricksets and chisels. Levels are used to establish layouts and check stacked stone or masonry units. Among the more useful levels are a 4-ft. level for measuring slopes and grade, a torpedo level for checking individual units, and a line level for setting layout strings.

From left to right: a wheelbarrow (A) with at least 5-cu.-ft. capacity is a necessity for transporting stones, loose fill, and other supplies; a corded circular saw with a masonry blade (B), preferably diamond-tipped, is used to score and cut stones and masonry units; landscape paint (C) is used to outline projects. The paint can be delivered with the can held upside down and it degrades harmlessly.

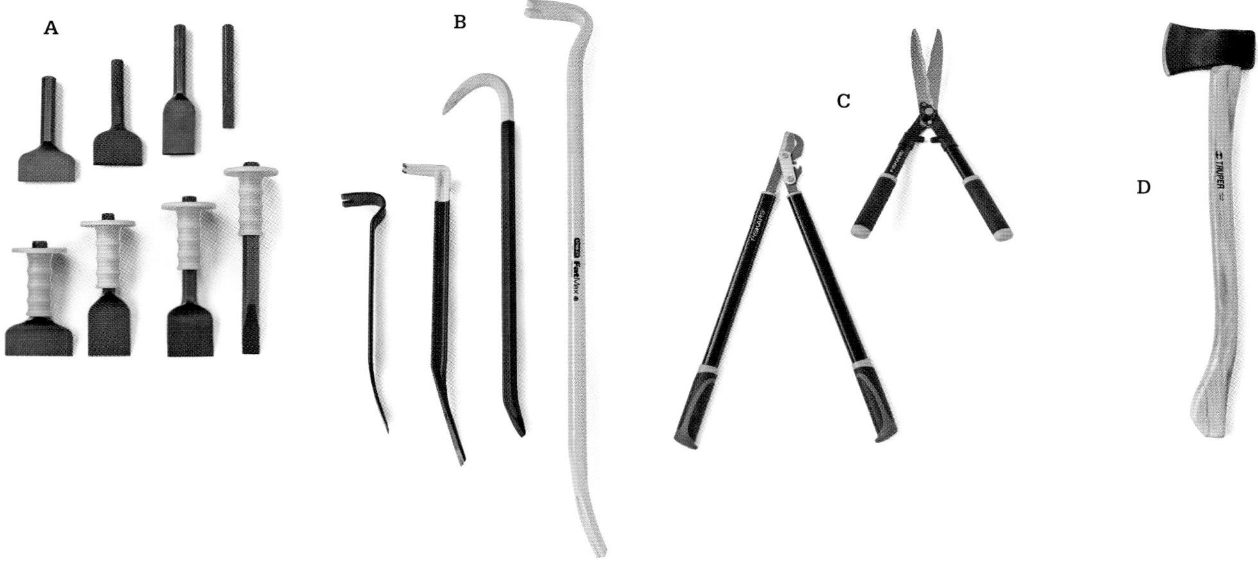

From left to right: stone chisels and bricksets (A) are struck with a maul to fracture or dress stones and masonry units; prybars and wrecking bars (B) are used to reposition larger stones; loppers and pruning shears (C) are used to trim back overgrown plants; an axe (D) can be used to cut through roots in an excavation area or to cut down small tress or shrubs.

Tool Tip: Safety Gear ▸

Cutting and hauling stone calls for special safety equipment, including:

- Protective knee pads
- Steel-toe work boots
- Hard hat
- Particle mask
- Sturdy gloves
- Eye and ear protection

Types & Forms of Stone

When choosing stone, you'll need to decide what type and what form to use. If you're shopping at a stone supply yard, you'll also find a wide range of shades and textures.

The most common types of stone for outdoor construction are shown below. In addition to a distinctive look, each type has a specific durability and workability to consider. If you expect to do a lot of splitting, ask your local stone supplier to help you select stone that splits easily. If you're laying a walk, select stone that holds up well under foot traffic. Cost, of course, is also a factor. Other things being equal, you will find that stone native to your area is the most affordable.

A stone's form can be thought of as its shape or cut. Common forms (right) include flagstone, fieldstone, marble, ashlar, veneer, and cobblestone. Some stone is uncut because its natural shape lends itself to certain types of construction. Stone is cut thin for use as facing (veneer) and wall caps (capstone). Often, the project dictates the form of stone to use. For example, most arches require stone with smooth, roughly square sides, such as ashlar, that can be laid up with very thin mortar joints.

Once you've determined the type and form of stone for your project, you can browse the wide range of shades and textures available and decide what best complements the look and feel of your yard.

Note: You may find that in your area different terms are used for various types of stone. Ask your supply yard staff to help you.

Limestone is a heavy stone, moderately easy to cut, medium to high strength, used in garden walls, rock gardens, walks, steps, and patios. Major U.S. sources: Indiana, Wisconsin, Kansas, and Texas.

Granite is a dense, heavy stone, difficult to cut, used for paving walks and building steps and walls; the most widely used building stone. Major U.S. sources: Massachusetts, Georgia, Minnesota, North Carolina, South Dakota, and Vermont.

Sandstone is a relatively lightweight stone available in "soft" and "dense" varieties and a wide range of colors. Soft sandstone is easier to cut but also lower in strength; used in garden walls, especially in frost-free climates. Major U.S. sources: New York, Arizona, Ohio, and Pennsylvania.

Slate is a fine, medium-weight stone that is soft and easy to cut but low in strength; too brittle for wall construction but a popular choice for walks, steps, and patios; colors vary widely from region to region. Major U.S. sources: Pennsylvania, Virginia, Vermont, Maine, New York, and Georgia.

Flagstone is large slabs of quarried stone cut into pieces up to 3" thick; it is used in walks, steps, and patios. Pieces smaller than 16" sq. are often called steppers.

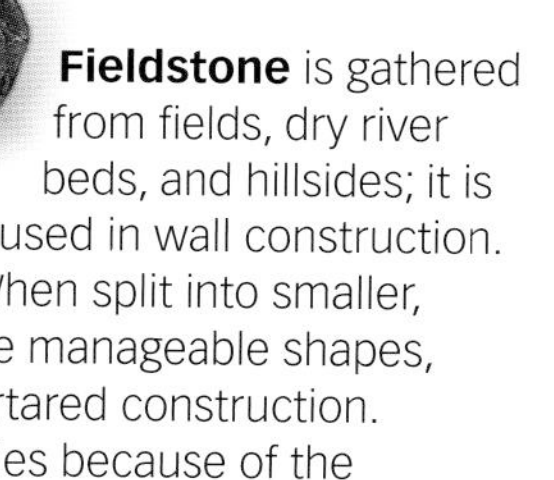

Fieldstone is gathered from fields, dry river beds, and hillsides; it is used in wall construction. When split into smaller, more manageable shapes, fieldstone is often used in mortared construction. Called river rock by some quarries because of the river-bed origin of some fieldstone.

Ashlar is quarried stone smooth-cut into large blocks ideal for creating clean lines with thin mortar joints.

Rubble is irregular pieces of quarried stone, usually with one split or finished face; widely used in wall construction.

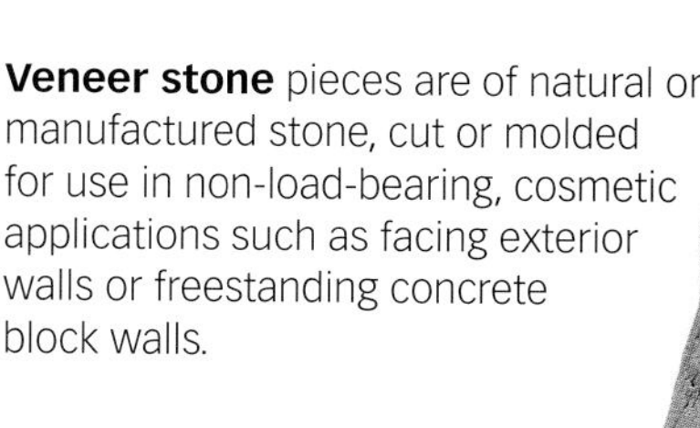

Veneer stone pieces are of natural or manufactured stone, cut or molded for use in non-load-bearing, cosmetic applications such as facing exterior walls or freestanding concrete block walls.

Cobblestone is small cuts of quarried stone or fieldstone; it is used in walks and paths.

Cutting Stone

Designate an area for cutting and shaping, preferably a grassy spot that will absorb the shock of heavy tools striking stone. This is your "cutting zone." It is important to keep children out of the dangerous area where stone chips could fly. You may use sandbags to anchor rounded stones while cutting. Or, you can build a banker to absorb shock (see below). A banker resembles a small sandbox, and you construct it with two layers of stacked 2 × 2s, forming a frame. Sandwich a piece of ¾-inch-thick plywood between the two layers. Pour sand on top of the plywood. You can set stone in the sand while cutting and shaping. If you prefer to stand while cutting, build up your banker by laying a foundation of stacked concrete block.

Now that your cutting surface is in order, collect all necessary tools and materials. The type of stone will dictate this. Always wear protective goggles and gloves while cutting, and if you are creating dust, wear a nuisance-rated particle mask (using wet-cutting techniques is a good way to limit the dust.)

Cutting stone calls for heavy-duty tools: a pitching chisel for long clean cuts; a pointing chisel for removing small bumps; a basic stone chisel; a sturdy maul; a sledgehammer; and a mason's hammer, which has a pick at one end that is helpful for breaking off small chips. For a circular saw, use a masonry blade (preferrably diamond-tipped) designed for the material you are cutting. Hard material like marble and concrete will require a different blade than softer stones like flagstone and limestone. Along with the hardware, you'll need a pipe or 2 × 4 for cutting flagstone. Also, keep on hand chalk or a crayon for marking cut lines.

A banker is a sand-filled wood box that provides a shock-absorbent surface for cutting stone.

Breaking stone is a simple process, but it requires a lot of practice to be done well. A stone chisel, a maul, and a soft surface are the primary tools you'll need.

Lifting and Moving Stone ▸

Many of the projects in this book require large, heavy stones. Even small stones can cause injury to your back if you don't lift and move them properly. Always support your back with lifting belts. Always bend at your knees when you lift stone. If you can't straighten up, the stone is too heavy to lift by yourself.

Other helpful stone-moving tools are ramps and simple towing devices, such as chains. When stacking interlocking block to construct a retaining wall, you may need a "helper" to lift stones as you build up some height. You can use 2 × 4s as ramps, placing a couple of them side-by-side to accommodate larger blocks. Using stone as a support underneath a ramp (2 × 4s), angle the ramp from the ground to the retaining wall. From a squatting position, push the stone up the ramp using your knees, not your back. You can also use a come-along tool to drag heavy stones. Gloves are a good idea, too.

Tips for Cutting Stone ▸

Laying stones works best when the sides (including the top and bottom) are roughly square. If a side is sharply skewed, score and split it with a pitching chisel, and chip off smaller peaks with a pointing chisel or mason's hammer. *Remember: A stone should sit flat on its bottom or top side without much rocking.*

"Dress" a stone using a pointing chisel and maul to remove jagged edges or undesirable bumps. Position the chisel at a 30 to 45° angle at the base of the piece to be removed. Tap lightly all around the break line, and then more forcefully, to chip off the piece. Position the chisel carefully before each blow with the maul.

Choosing Your Chisel

A pointing chisel is used to clean up edges and surfaces by knocking off small chunks and ridges.

A pitching chisel has a relatively wide blade for making long, clean cuts.

A basic stone chisel can handle a variety of stonecutting tasks, including cleaving stones in two.

Cutting Stone with a Circular Saw ▸

A circular saw lets you precut stones with broad surfaces with greater control and accuracy than most people can achieve with a chisel. It's a noisy tool, so wear ear plugs, along with a dust mask and safety goggles. Install a toothless masonry blade on your saw and start out with the blade set to cut ⅛" deep. Make sure the blade is designed for the material you're cutting. Some masonry blades are designed for hard materials like concrete, marble, and granite. Others are for soft materials, like concrete block, brick, flagstone, and limestone. Wet the stone before cutting to help control dust, and then make three passes, setting the blade ⅛" deeper with each pass. Repeat the process on the other side. A thin piece of wood under the saw protects the saw foot from rough masonry surfaces. *Remember: Always use a GFCI outlet or extension cord when using power tools outdoors.*

How to Cut Fieldstone

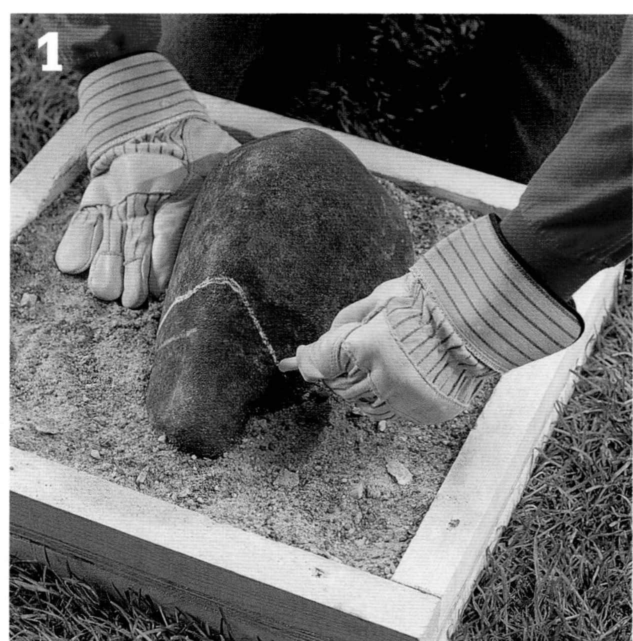

Place the stone on a banker or prop it with sandbags and mark with chalk or a crayon all the way around the stone, indicating where you want it to split. If possible, use the natural fissures in the stone as cutting lines.

Score along the line using moderate blows with a chisel and maul, and then strike solidly along the score line with a pitching chisel to split the stone. Dress the stone with a pointing chisel.

How to Cut Flagstone

Trying to split a large flagstone in half can lead to many unpredicted breaks. For best results, chip off small sections at a time. Mark the stone on both sides with chalk or a crayon, indicating where you want it to split. If there is a fissure nearby, mark your line there since that is probably where the stone will break naturally.

Score along the line on the back side of the stone (the side that won't be exposed) by moving a chisel along the line and striking it with moderate blows with a maul. *Option: If you have a lot of cutting to do, reduce hammering fatigue by using a circular saw to score the stones and a maul and chisel to split them. Keep the stone wet during cutting with a circular saw to reduce dust.*

Turn the stone over, place a pipe or 2 × 4 directly under the chalk line, and then strike forcefully with the maul on the end of the portion to be removed.

Option: If a paving stone looks too big compared to other stones in your path, simply set the stone in place and strike a heavy blow to the center with a sledge hammer. It should break into several usable pieces.

Laying Stone

The methods of laying stone are as varied as the stone masons who practice the craft. But all of them would agree on a few general principles:

- Thinner joints are stronger joints. Whether you are using mortar or dry-laying stone, the more contact between stones, the more resistance to any one stone dislodging.
- Tie stones are essential in vertical structures, such as walls or pillars. These long stones span at least two-thirds of the width of the structure, tying together the shorter stones around them.
- When working with mortar, most stone masons point their joints deep for aesthetic reasons. The less mortar is visible, the more the stone itself is emphasized.
- Long vertical joints, or head joints, are weak spots in a wall. Close the vertical joints by overlapping them with stones in the next course, similar to a running-bond pattern in a brick or block wall.
- The sides of a stone wall should have an inward slope (called batter) for maximum strength. This is especially important with dry-laid stone. Mortared walls need less batter.

Thin joints are the strongest. When working with mortar, joints should be ½ to 1" thick. Mortar is not intended to create gaps between stones but to fill the inevitable gaps and strengthen the bonds between stones. Wiggle a stone once it is in place to get it as close as possible to adjoining stones.

Blend large and small stones in walks or in vertical structures to achieve the most natural appearance. In addition to enhancing visual appeal, long stones in a walk act like the tie stones in a wall, adding strength by bonding with other stones.

Place uneven stone surfaces down and dig out the soil underneath until the stone lies flat. Use the same approach in the bottom course of a dry-laid wall, only make sure stones at the base of a wall slope toward the center of the trench.

Tips for Laying Stone Walls ▸

Tie stones are long stones that span most of the width of a wall, tying together the shorter stones and increasing the wall's strength. As a guide, figure that 20% of the stones in a structure should be tie stones.

A shiner is the opposite of a tie stone—a flat stone on the side of a wall that contributes little in terms of strength. A shiner may be necessary when no other stone will fit in a space. Use shiners as seldom as possible, and use tie stones nearby to compensate.

Lay stones in horizontal courses, where possible (a technique called ashlar construction). If necessary, stack two to three thin stones to match the thickness of adjoining stones.

With irregular stone, such as untrimmed rubble or fieldstone, building course by course is difficult. Instead, place stones as needed to fill gaps and to overlap the vertical joints.

Use a batter gauge and level to lay up dry stone structures so the sides angle inward. Angle the sides of a wall 1" for every 2 ft. of height—less for ashlar and freestanding walls, twice as much for round stone and retaining walls.

Dry Stone Wall

Stone walls are beautiful, long-lasting structures that are surprisingly easy to build provided you plan carefully. A low stone wall can be constructed without mortar using a centuries-old method known as dry laying. With this technique, the wall is actually formed by two separate stacks that lean together slightly. The position and weight of the two stacks support each other, forming a single, sturdy wall. A dry stone wall can be built to any length, but its width must be at least half of its height.

You can purchase stone for this project from a quarry or stone supplier, where different sizes, shapes, and colors of stone are sold, priced by the ton. The quarry or stone center can also sell you Type M mortar—necessary for bonding the capstones to the top of the wall.

Building dry stone walls requires patience and a fair amount of physical effort. The stones must be sorted by size and shape. You'll probably also need to shape some of the stones to achieve consistent spacing and a general appearance that appeals to you.

To shape a stone, score it using a circular saw outfitted with a masonry blade. Place a mason's chisel on the score line and strike with a maul until the stone breaks. Wear safety glasses when using stonecutting tools.

Tools & Materials ▸

- Mason's string and stakes
- Compactable gravel
- Ashlar stone
- Capstones
- Mortar mix
- Trowel
- Stiff-bristle brush
- Work gloves

It is easiest to build a dry stone wall with ashlar—stone that has been split into roughly rectangular blocks. Ashlar stone is stacked in the same running-bond pattern used in brick wall construction; each stone overlaps a joint in the previous course. This technique avoids long vertical joints, resulting in a wall that is attractive and also strong.

How to Build a Dry Stone Wall

Lay out the wall site using stakes and mason's string. Dig a 6"-deep trench that extends 6" beyond the wall on all sides. Add a 4" crushed stone sub-base to the trench, creating a "V" shape by sloping the sub-base so the center is about 2" deeper than the edges.

Select appropriate stones and lay the first course. Place pairs of stones side by side, flush with the edges of the trench and sloping toward the center. Use stones of similar height; position uneven sides face down. Fill any gaps between the shaping stones with small filler stones.

Lay the next course, staggering the joints. Use pairs of stones of varying lengths to offset the center joint. Alternate stone length, and keep the height even, stacking pairs of thin stones if necessary to maintain consistent height. Place filler stones in the gaps.

Every other course, place a tie stone every 3 ft. You may need to split the tie stones to length. Check the wall periodically for level.

Mortar the capstones to the top of the wall, keeping the mortar at least 6" from the edges so it's not visible. Push the capstones together and mortar the cracks in between. Brush off dried excess mortar with a stiff-bristle brush.

Mortared Stone Wall

The mortared stone wall is a classic that brings structure and appeal to any yard or garden. Square-hewn ashlar and bluestone are the easiest to build with, though fieldstone and rubble also work well and make attractive walls.

Because the mortar turns the wall into a monolithic structure that can crack and heave with a freeze-thaw cycle, a concrete footing is required for a mortared stone wall. To maintain strength in the wall, use the heaviest, thickest stones for the base of the wall and thinner, flatter stones for the cap.

As you plan the wall layout, install tie stones—stones that span the width of the wall (page 183)—about every three feet, staggered through the courses both vertically and horizontally throughout the wall. Use the squarest, flattest stones to build the "leads," or ends of the wall first, and then fill the middle courses. Plan for joints around one inch thick and make sure joints in successive courses do not line up. Follow this rule of thumb: Cover joints below with a full stone above; locate joints above over a full stone below.

Laying a mortared stone wall is labor-intensive but satisfying work. Make sure to work safely and enlist friends to help with the heavy lifting.

Tools & Materials ▸

- Tape measure
- Pencil
- Chalk line
- Small whisk broom
- Tools for mixing mortar
- Maul
- Stone chisel
- Pitching chisel
- Trowel
- Jointing tool
- Line level
- Sponge
- Garden hose
- Concrete materials for foundation
- Ashlar stone
- Type N or Type S mortar
- Stakes and mason's line
- Scrap wood
- Muriatic acid
- Bucket of water
- Sponge
- Eye protection and work gloves

A mortared stone wall made from ashlar adds structure and classic appeal to your home landscape. Plan carefully and enlist help to ease the building process.

How to Build a Mortared Stone Wall

Pour a footing for the wall and allow it to cure for one week (pages 46 to 49). Measure and mark the wall location so it is centered on the footing. Snap chalk lines along the length of the footing for both the front and the back faces of the wall. Lay out corners using the 3-4-5 right angle method.

Dry-lay the entire first course. Starting with a tie stone at each end, arrange stones in two rows along the chalk lines with joints about 1" thick. Use smaller stones to fill the center of the wall. Use larger, heavier stones in the base and lower courses. Place additional tie stones approximately every 3 ft. Trim stones as needed.

Mix a stiff batch of Type N or Type S mortar, following the manufacturer's directions (pages 28 to 29). Starting at an end or corner, set aside some of the stone and brush off the foundation. Spread an even, 2" thick layer of mortar onto the foundation, about ½" from the chalk lines—the mortar will squeeze out a little.

Firmly press the first tie stone into the mortar so it is aligned with the chalk lines and relatively level. Tap the top of the stone with the handle of the trowel to set it. Continue to lay stones along each chalk line, working to the opposite end of the wall.

(continued)

After installing the entire first course, fill voids along the center of the wall that are larger than 2" with smaller rubble. Fill the remaining spaces and joints with mortar, using the trowel.

As you work, rake the joints using a scrap of wood to a depth of ½"; raking joints highlights the stones rather than the mortared joints. After raking, use a whisk broom to even the mortar in the joints.

Variation: You can also tool joints for a cleaner, tighter mortared joint. Tool joints when your thumb can leave an imprint in the mortar without removing any of it.

Drive stakes at the each end of the wall and align a mason's line with the face of the wall. Use a line level to level the string at the height of the next course. Build up each end of the wall, called the "leads," making sure to stagger the joints between courses. Check the leads with a 4-ft. level on each wall face to make sure it is plumb.

If heavy stones push out too much mortar, use wood wedges cut from scrap to hold the stone in place. Once the mortar sets up, remove the wedges and fill the voids with fresh mortar.

Removing Mortar ▸

Have a bucket of water and a sponge handy in case mortar oozes or spills onto the face of the stone. Wipe mortar away immediately before it can harden.

Fill the middle courses between the leads by first dry laying stones for placement and then mortaring them in place. Install tie stones about every 3 feet, both vertically and horizontally, staggering their position in each course. Make sure joints in successive courses do not fall in alignment.

Install cap stones by pressing flat stones that span the width of the wall into a mortar bed. Do not rake the joints, but clean off excess mortar with the trowel and clean excess mortar from the surface of the stones using a damp sponge.

Allow the wall to cure for one week, and then clean it using a solution of 1 part muriatic acid and 10 parts water. Wet the wall using a garden hose, apply the acid solution, and then immediately rinse with plenty of clean, clear water. Always wear goggles, long sleeves and pants, and heavy rubber gloves when using acids.

Stone Retaining Wall

Rough-cut wall stones may be dry stacked (without mortar) into retaining walls, garden walls, and other stonescape features. Dry-stack walls are able to move and shift with the frost, and they also drain well so they don't require deep footings and drainage tiles. Unlike fieldstone and boulder walls, short wall-stone walls can be just a single stone thick.

In the project featured here, we use rough-split limestone blocks about eight by four inches thick and in varying lengths. Walls like this may be built up to three feet tall, but keep them shorter if you can, to be safe. Building multiple short walls is often a more effective way to manage a slope than to build one taller wall. Called terracing, this practice requires some planning. Ideally, the flat ground between pairs of walls will be approximately the uniform size.

A dry-laid natural stone retaining wall is a very organic-looking structure compared to interlocking block retaining walls (see page 134). One way to exploit the natural look is to plant some of your favorite stone-garden perennials in the joints as you build the wall(s). Usually one plant or a cluster of three will add interest to a wall without suffocating it in vegetation or compromising its stability. Avoid plants that get very large or develop thick, woody roots or stems that may compromise the stability of the wall.

A well-built retaining wall has a slight lean, called a batter, back into the slope. It has a solid base and the bottom course is dug in behind the lower terrace. Drainage gravel can help keep the soil from turning to mud, which will slump and press against the wall.

The same basic techniques used to stack natural stone in a retaining wall may be used for building a short garden wall as well. Obviously, there is no need for drainage allowances or wall returns in a garden wall. Simply prepare a base similar to the one shown here and begin stacking. The wall will look best if it wanders and meanders a bit. Unless you're building a very short wall (less than 18 inches), use two parallel courses that lean against one another for the basic construction. Cap it with flat capstones that run the full width of the wall (see page 192).

Tools & Materials ▸

- Goggles, gloves, steel-toe boots
- Mattock with pick
- Hatchet or loppers
- Spades
- Measuring tape
- Mason's string
- Line level
- Stakes
- Hand maul
- Garden rake
- Torpedo level
- Straight 2 × 4
- Hand tamper
- Compactable gravel
- Ashlar wall stone
- Drainage gravel
- Landscape fabric
- Block-and-stone adhesive
- Caulk gun

A natural stone retaining wall not only adds a stunning framework to your landscape, but it also lends a practical hand to prevent hillsides and slopes from deteriorating over time.

Cross Sections: Stone Retaining Walls

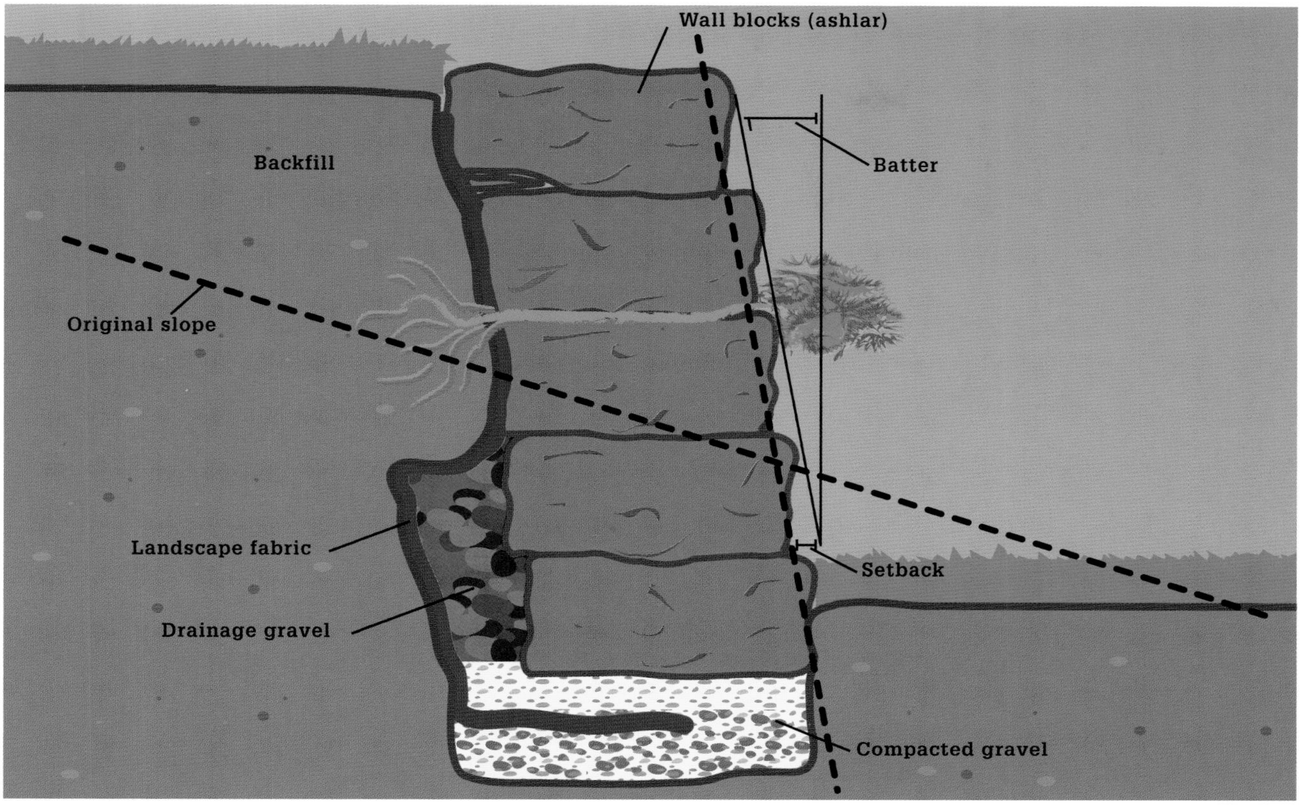

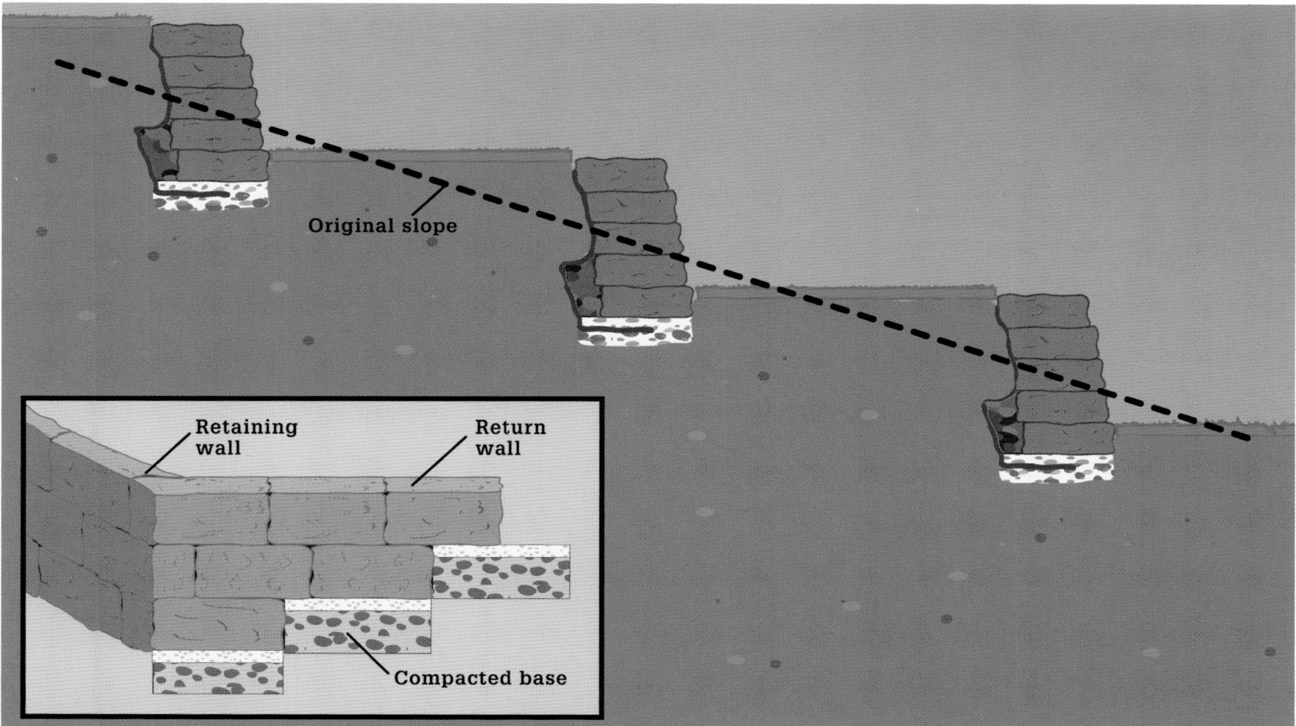

A stone retaining wall breaks up a slope to neat flat lawn areas that are more usable (top). A series of walls and terraces (bottom) break up larger slopes. Short return walls (inset) create transitions to the yard.

How to How to Build a Stone Retaining Wall

Dig into the slope to create a trench for the first wall. Reserve the soil you remove nearby—you'll want to backfill with it when the wall is done.

Level the bottom of the trench and measure to make sure you've excavated deeply enough.

After compacting a base, cover the trench and hill slope with landscape fabric, and then pour and level a 1" (3 cm) layer of coarse sand.

Place the first course of stones in rough position. Run a level mason's string at the average height of the stones.

Add or remove gravel under each stone to bring the front edges level with the mason's string.

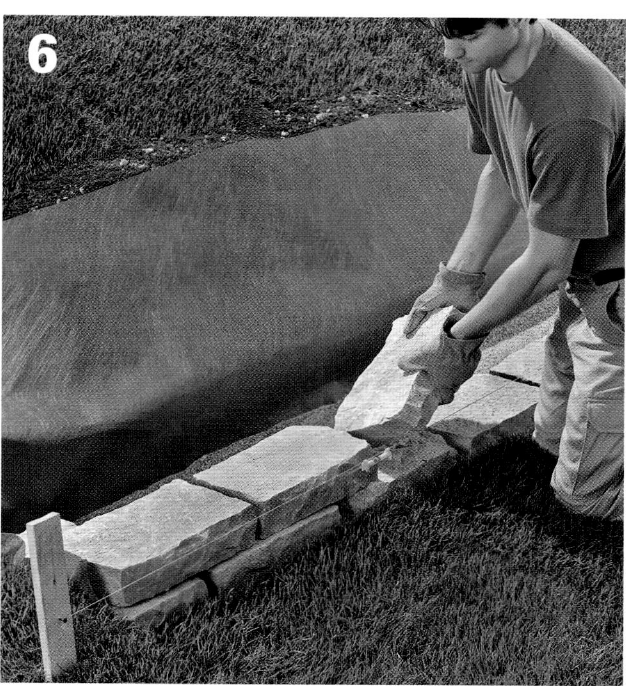

Begin the second course with a longer stone on each end so the vertical gaps between stones are staggered over the first course.

Finish out the second course. Use shards and chips of stone as shims where needed to stabilize the stones. Check to make sure the ½" (3 mm) setback is followed.

Finish setting the return stones in the second course, making adjustments as needed for the return to be level.

(continued)

Backfill behind the wall with river rock or another good drainage rock.

Fold the landscape fabric over the drainage rock (the main job of the fabric is to keep soil from migrating into the drainage rock and out the wall) and backfill behind it with soil to level the ground.

Trim the landscape fabric just behind the back of the wall, near the top.

Finish the wall by capping it off with some of your nicer, long flat stones. Bond them with block-and-stone adhesive.

Level off the soil behind the wall with a garden rake. Add additional walls if you are terracing.

Planting Your Retaining Wall ▸

Natural stone retaining walls look quite lovely in their own right. However, you can enhance the effect by making some well-chosen planting choices for the wall itself. You can plan for this in the wall construction by leaving an extra wide gap between two stones in one of the courses and then planting in the gap. Or you can replace a stone in the wall with a shorter one, also creating a gap. To plant a gap, cut the fabric and set a good-size, bare-root perennial of an appropriate species to the bottom of this joint. Fan out the roots over the soil and use sphagnum moss to plug up the gaps in the wall around plants. Adhere the stone in the next course that bridges the gap with block-and-stone adhesive. Keep plants well watered until established. Eventually, the plant roots will hold the soil instead of the moss.

Set plants in natural-looking clusters of the same species. Do not suffocate the wall with too many plants.

Arroyo (Dry Streambed)

An arroyo is a dry streambed or watercourse in an arid climate that directs water runoff on the rare occasions when there is a downfall. In a home landscape an arroyo may be used for purely decorative purposes, with the placement of stones evoking water where the real thing is scarce. Or it may serve a vital water-management function, directing storm runoff away from building foundations to areas where it may percolate into the ground and irrigate plants, creating a great spot for a rain garden. This water management function is becoming more important as municipalities struggle with an overload of storm sewer water, which can degrade water quality in rivers and lakes. Some communities now offer tax incentives to homeowners who keep water out of the street.

When designing your dry streambed, keep it natural and practical. Use local stone that's arranged as it would be found in a natural stream. Take a field trip to an area containing natural streams and make some observations. Note how quickly the water depth drops at the outside of bends where only larger stones can withstand the current. By the same token, note how gradually the water level drops at the inside of broad bends where water movement is slow. Place smaller river-rock gravel here, as it would accumulate in a natural stream.

Large heavy stones with flat tops may serve as step stones, allowing paths to cross or even follow dry stream beds.

The most important design standard with dry streambeds is to avoid regularity. Stones are never spaced evenly in nature and nor should they be in your arroyo. If you dig a bed with consistent width, it will look like a canal or a drainage ditch, not a stream. And consider other yard elements and furnishings. For example, an arroyo presents a nice opportunity to add a landscape bridge or two to your yard.

An arroyo is a drainage swale lined with rocks that direct runoff water from a point of origin, such as a gutter downspout, to a destination, such as a sewer drain or a rain garden.

Important: Contact your local government before deliberately routing water toward a storm sewer; this may be illegal.

Tools & Materials ▸

Landscape paint
Carpenter's level
Spades
Garden rake
Wheelbarrow
Landscape fabric
6-mil black plastic
Mulch
8" (20 cm)-thick steppers
6 to 18" dia. (15 to 46 cm) river-rock boulders
¾ to 2" (19 to 51 mm) river rock
Native grasses or other perennials for banks
Eye protection and work gloves

How to Build an Arroyo

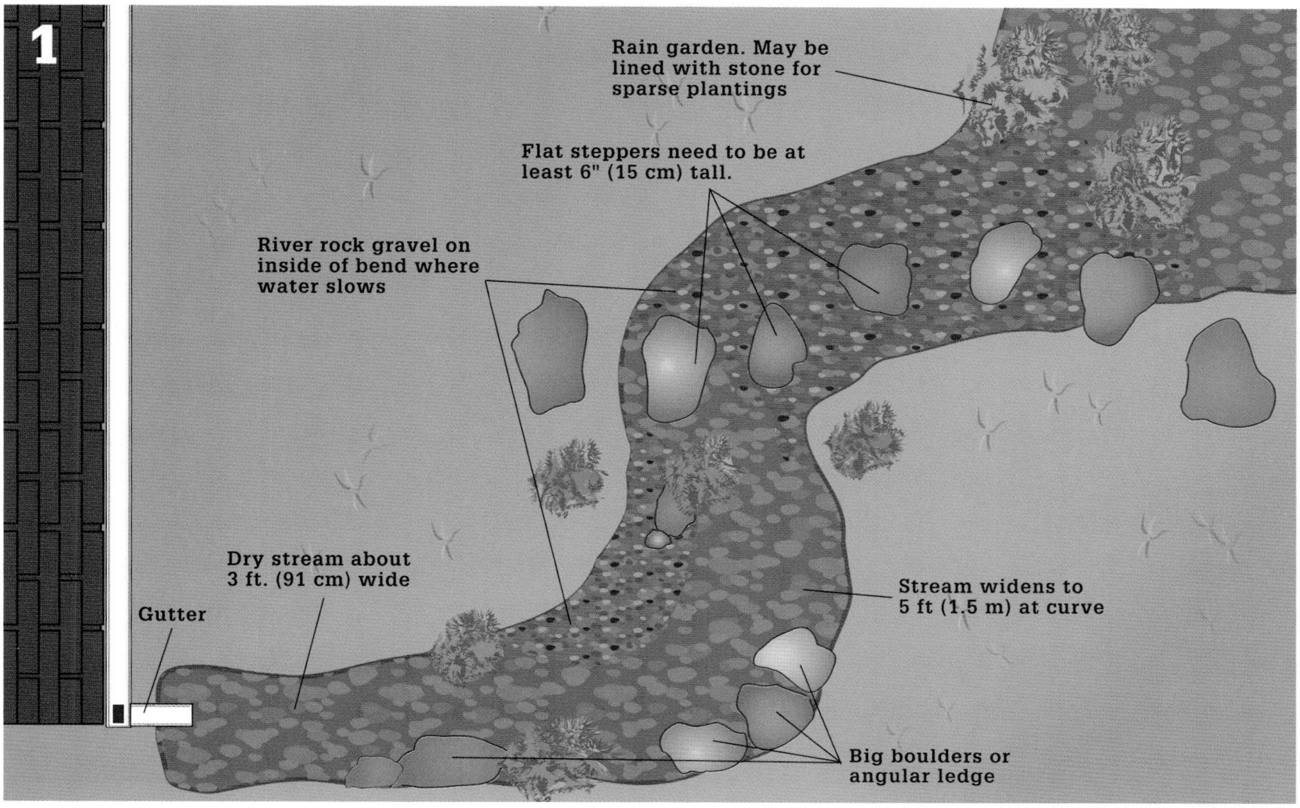

1

Rain garden. May be lined with stone for sparse plantings

Flat steppers need to be at least 6" (15 cm) tall.

River rock gravel on inside of bend where water slows

Dry stream about 3 ft. (91 cm) wide

Gutter

Stream widens to 5 ft (1.5 m) at curve

Big boulders or angular ledge

Create a plan for the arroyo. The best designs have a very natural shape and a rock distribution strategy that mimics the look of a stream. Arrange a series of flat steppers at some point to create a bridge.

Lay out the dry stream bed, following the native topography of your yard as much as possible. Mark the borders and then step back and review it from several perspectives.

Excavate the soil to a depth of at least 12" (30 cm) in the arroyo area. Use the soil you dig up to embellish or repair your yard.

(continued)

Widen the arroyo in selected areas to add interest. Rake and smooth out the soil in the project area.

Install an underlayment of landscape fabric over the entire dry streambed. Keep the fabric loose so you have room to manipulate it later if the need arises.

Set larger boulders at outside bends in the arroyo. Imagine that there is a current to help you visualize where the individual stones could naturally end up.

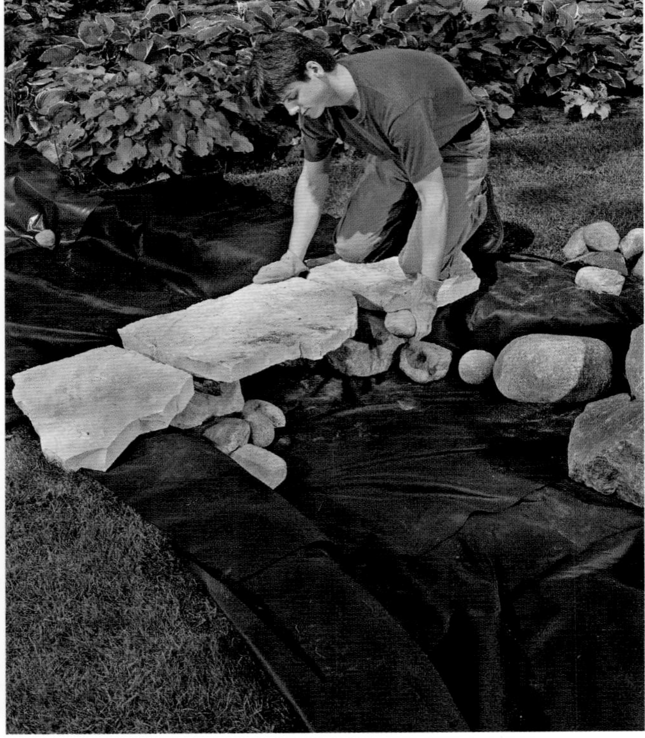

Place flagstone steppers or boulders with relatively flat surfaces in a stepping-stone pattern to make a pathway across the arroyo (left photo). Alternately, create a "bridge" in an area where you're likely to be walking (right photo).

8

Add more stones, including steppers and medium-size landscape boulders. Use smaller aggregate to create the stream bed, filling in, around, but not covering, the larger rocks.

Dress up your new arroyo by planting native grasses and perennials around its banks.

What is a Rain Garden? ▸

A rain garden is simply a shallow, wide depression at least ten feet away from a basement foundation that collects storm water runoff. Rain gardens are planted with native flood-tolerant plants and typically hold water for only hours after rainfall. Check your local garden center or Extension Service to find details about creating rain gardens in your area.

Stone Firepit

A firepit is a backyard focal point and gathering spot. The one featured here is constructed around a metal liner, which will keep the fire pit walls from overheating and cracking if cooled suddenly by rain or a bucket of water. The liner here is a section of 36-inch-diameter corrugated culvert pipe. Check local codes for stipulations on pit area size. Many codes require a 20-foot-diameter pit area.

Ashlar wall stones add character to the fire pit walls, but you can use any type of stone, including cast concrete retaining wall blocks. You'll want to prep the base for the seating area as you dig the fire pit to be sure both rest on the same level plane.

Tools & Materials ▸

- Wheelbarrow
- Landscape paint
- String and stakes
- Spades
- Metal pipe
- Landscape edging
- Level
- Garden rake
- Plate vibrator
- Metal firepit liner
- Compactable gravel
- Top-dressing rock (trap rock)
- Wall stones
- Eye protection and work gloves

Some pointers to consider when using your fire pit include: 1) Make sure there are no bans or restrictions in effect; 2) Evaluate wind conditions and avoid building a fire if winds are heavy and/or blowing toward your home; 3) Keep shovels, sand, water, and a fire extinguisher nearby; 4) Extinguish fire with water and never leave the fire pit unattended.

Cross Section: Firepit

Wall blocks
(ashlar or interlocking)
Liner
(culvert pipe)
Nonflammable
ground cover
2"
2"
4"
Compacted gravel

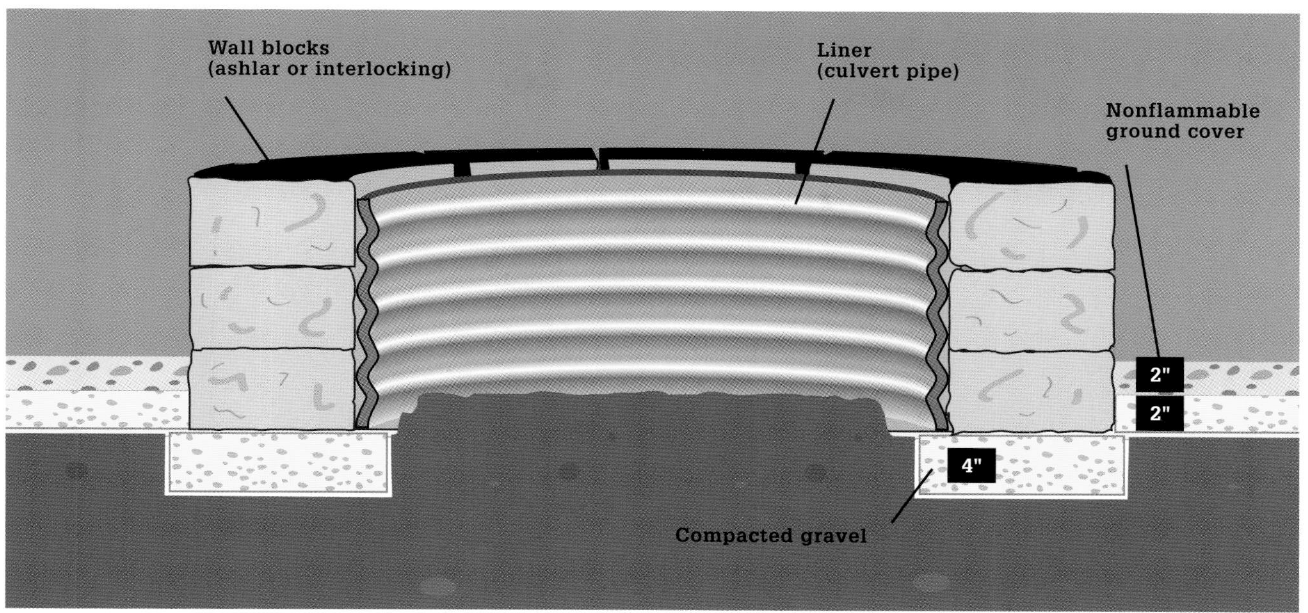

Plan View: Firepit

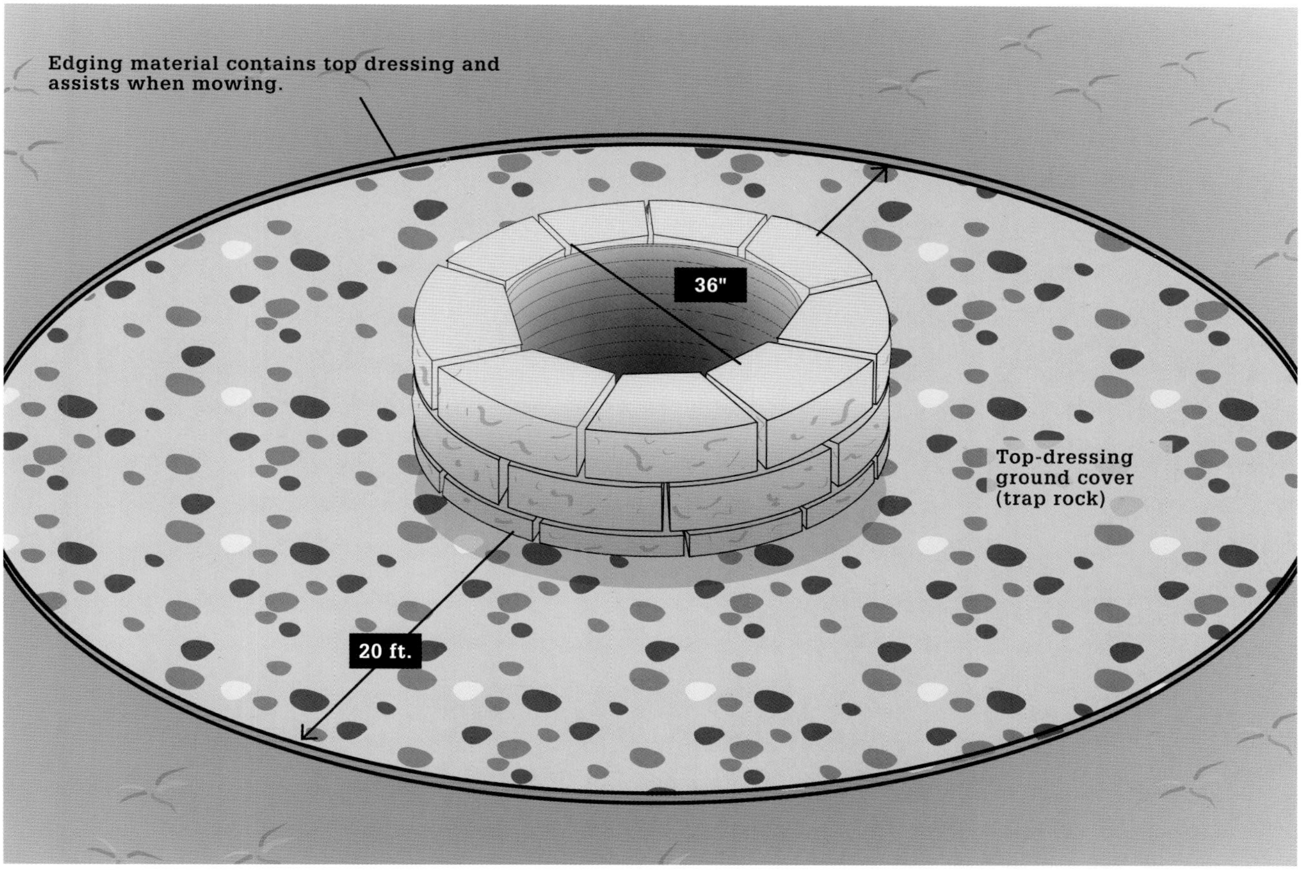

Edging material contains top dressing and
assists when mowing.
36"
Top-dressing
ground cover
(trap rock)
20 ft.

How to Build a Firepit

Outline the location for your firepit and the firepit safety area by drawing concentric circles with landscape paint using a string and pole for guidance.

Remove a 4"-deep layer of sod and dirt in the firepit and safety areas (the depth of the excavation depends on what materials you're installing in the safety zone.)

Dig a 4"-deep trench for the perimeter stones that will ring the pit liner.

Fill the trench for the perimeter stones with compactable gravel and tamp thoroughly. Then scatter gravel to within 2½" of the paver edging top throughout the project area. It is not necessary to tamp this layer at this time.

Place your metal fire ring so it is level on the gravel layer and centered around the center pipe.

Arrange the first course of wall blocks around the fire ring. Keep gaps even and check with a level, adding or removing gravel as needed.

Install the second course of retaining wall block, taking care to evenly stagger the vertical joints on the first and second courses. Add the remaining courses.

Compact the compactable gravel in the seating/safety area using a rental plate vibrator.

Place and compact a layer of top-dressing rock in the seating/safety area to complete the firepit.

Mortared Flagstone Patio

The stately flagstone patio can be sandset or mortared using lightly trimmed stone or naturally irregular stone shapes. If you're sandsetting, which is a good idea in colder climates, you'll have the best luck if you cut the stones so they are as large as possible and have straight edges. Small stones don't provide much bearing when they're resting on the sand. But they can be a very effective part of the design if you embed them into mortar, as we do here.

You can install a mortared flagstone patio over an old concrete patio if it is reasonably good repair or you can pour a new concrete base that's at least two inches thick with five inches of compacted rock below for drainage. If you are sandsetting, you'll want a layer of coarse sand that's at least one or two inches thick over a well-compacted base of gravel (see the sandset flagstone patio project starting on page 226 for more information).

Tools & Materials ▸

- Paint roller with extension pole
- Pencil
- Small whisk broom
- Tools for mixing mortar
- Shovel
- Maul
- Stone chisel
- Pitching chisel
- 4-ft. level
- Rubber mallet
- Eye protection and work gloves
- Trowel
- Straight 2 × 4 stud
- Grout bag
- Jointing tool
- Sponge
- Garden hose
- Concrete bonding agent
- Flagstone stone
- Type N or Type S mortar
- Acrylic fortifier
- Stone sealer

The flagstone patio is a classic element of modern landscape design. Bluestone (above) is one of the most popular types but may not be available in all areas since specific types vary by region.

How to Build a Mortared Flagstone Patio

Thoroughly clean the concrete slab. While the slab doesn't need to be in perfect condition, it does need to be sound. Repair large cracks or holes. After repairs have cured, apply a latex bonding agent to the patio surface, following the manufacturer's instructions.

Once the bonding agent has set up per the manufacturer's recommendations, dry lay stones on the patio to determine an appealing layout. Work from the center outward and evenly distribute large stones and smaller ones, with ½ to 1" joints between them.

Cut stones to size as needed. Mark the cutting line with chalk, and then cut the stone, following the techniques on page 179. At the sides of the slab, cut stones even with the edges to accommodate edging treatments.

Variation: For a more rustic appearance, allow stones to overhang the edges of the slab. Stones thicker in size can overhang as much as 6", provided that the slab supports no less than two-thirds of the stone. Thinner stones should not overhang more than 3". After stones are mortared in place, fill in beneath the overhanging stones with soil.

(continued)

Mix a stiff batch of Type N or Type S mortar, following the manufacturer's directions. Starting near the center of the patio, set aside some of the stone, maintaining their layout pattern. Spread a 2" thick layer of mortar onto the slab.

Firmly press the first large stone into the mortar, in its same position as in the layout. Tap the stone with a rubber mallet or the handle of the trowel to set it. Use a 4-ft. level and a scrap of 2 × 4 to check for level; make any necessary adjustments.

Using the first stone as a reference for the course height, continue to lay stones in mortar, working from the center of the slab to the edges. Maintain ½ to 1" joints.

As you work, check for level often, using a straight length of 2 × 4 and the 4-ft. level. Tap stones to make minor adjustments. Once you're done, let the mortar set up for a day or two before walking on it.

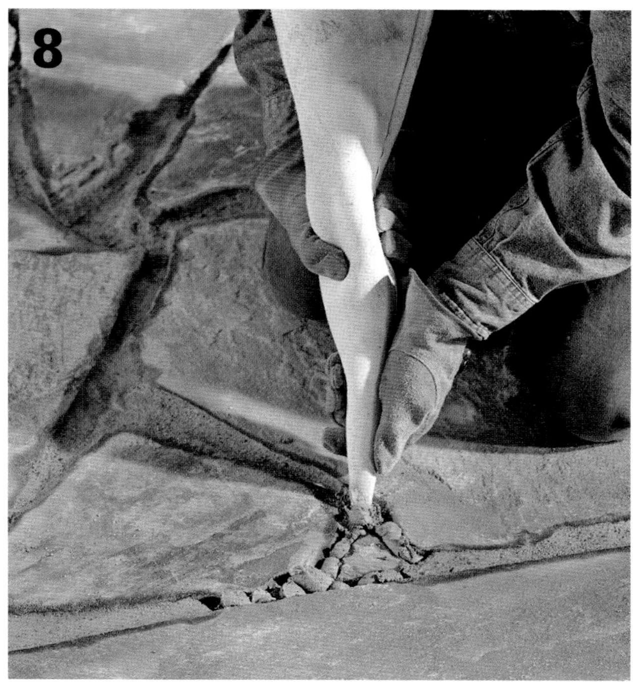

Use a grout bag to fill the joints with mortar (add acrylic fortifier to the mix to make the mortar more elastic). Do not overfill the joints. Pack loose gravel and small rocks into gaps first to conserve mortar and make stronger joints. Wipe up spilled mortar.

Once the mortar is stiff enough that your thumb leaves an impression without mortar sticking to it, rake the joints just enough so the mortar is even with the surface of the stone, so water cannot pool. Use a whisk broom to shape the mortar.

Allow the mortar to cure for a few days, and then clean the patio with clear water and a stiff bristle brush. After the mortar cures for a week, apply a stone sealer, following the manufacturer's instructions.

Flagstone Garden Steps

Flagstone garden steps are an ideal solution for managing low slopes. They consist of broad flagstone treads and blocky ashlar risers, commonly sold as wall stone. The risers are prepared with compactable gravel beds on which the flagstone treads rest. This project features flagstone and wall stone in their originally split state (as opposed to sawn).

The process of fitting stones together involves a lot of cutting and waste, so plan on purchasing 40 percent more stone material than your plans require. Choose stone with uniform thickness, if possible. Flagstone steps work best when you create the broadest possible treads. Think of them as a series of terraced patios.

Tools & Materials ▸

- Stakes
- Mason's string
- Landscape marking paint
- String level
- Straight 2 × 4
- Torpedo level
- 4-ft. level
- Measuring tape
- Spun-bonded landscape fabric
- Excavating tools
- Compactable gravel
- Coarse sand
- 3-lb. maul
- Hand tamper
- Wall stone
- Flagstones
- Block-and-stone adhesive
- Work gloves

Flagstone garden steps typically have a casual appearance that follows the shape of your yard naturally. Broad steppers are used for the treads with the front set on smaller steppers to level them and function as risers.

Cross Section: Garden Steps

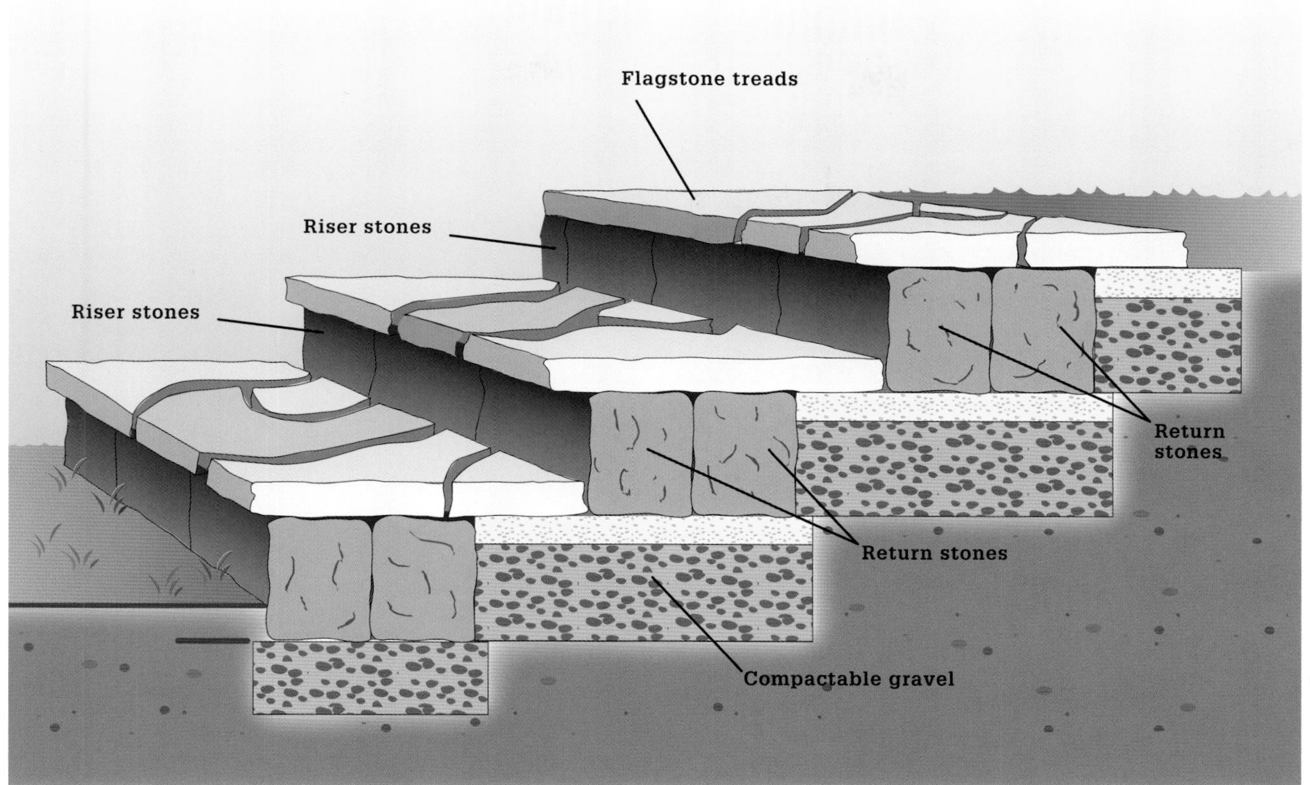

Stone Step Variations

Flagstone steps create a rustic pathway in a natural garden setting. They can be installed using the natural hill slope as a riser.

Concrete and natural stone create an elegant and uniform walkway for a gentle slope.

How to Build Flagstone Steps

Thread a line level onto a mason's string and tie the ends of the string to stakes at the top and bottom of the stair installation site. With the string level, measure the difference in distance from the string to the ground at the top and bottom of the steps to find the total run. See pages 58 to 59 for more help on designing steps.

Excavate for the first step and the stone walls risers and returns. Dig deep enough to accommodate 4" of compactable gravel and 1" of sand throughout. This means you'll be excavating a shallow area for the tread and a deeper U-shaped trench for the wall stones.

Pour a layer of compactable gravel into the U-shaped trench for the wall stones. Compact the gravel with a tamper or post and then top it off with another layer that should not be compacted.

Position the riser stones and the return stones in the trench and level them. Add or remove gravel as necessary and then rap them gently on the tops with a hand maul to set them. Use a wood block to protect the stones from the maul.

Line the area under the first tread with landscape fabric, drawing it up to cover the insides of the risers and returns. Add a layer of compactable gravel and tamp down to within 1" of risers and returns. Fill with sand and level with a 2 × 4. Slope gravel slightly from back to front for drainage.

Measure the step/run distance back from the face of the first risers and set a mason's line across the sand bed. Set the second course of risers and returns as you did the first, without digging risers on to the bottom (the bottom tread will reduce risers' effective height).

Begin laying out flagstone treads. First, position them like a puzzle to determine if cuts are necessary. Leave a consistent distance between stones. Allow steps to hang about 2" over risers.

Fill in gaps between larger stones by trimming smaller pieces to fit. Fill smaller stones near the back. Don't allow stones to touch one another when in place and do not cut stones too small. Ideally, each should be at least the size of a dinner plate.

(continued)

Pack wet sand underneath low areas and remove sand underneath high areas until all the flags on the tread are basically flat and even. Use a level as a guide.

Use thin pieces of broken stone as shims to raise wall stones to their required level. Make sure shims are sturdy enough that they won't flake apart easily. Use block-and-stone adhesive to hold the shims in place. Make sure there is no path for sand to wash out from beneath the treads. Do not use sand in place of shims to raise wall stones.

Continue adding steps and making your way up the slope. You shouldn't need to trench for risers, but you may need to move some dirt so you can pack it in and install the return stones. The bottom of the risers should be at the same height as the bottoms of the tread on the step below. The top step often will not require returns.

Fill the joints between stones with coarse sand to bind them together and for a more finished appearance. Granite sand works well for this purpose, or choose polymeric sand that resists wash-out better than regular builder's sand. Inspect steps regularly for the first few weeks and make adjustments to height of stones as needed.

Natural Stone Step Design ▸

Steps made using natural stones most frequently are found in locations where the setting is fairly wild, or at least very casual. Whenever you design and build steps, you should be aware of safety issues, but the fact is that in a rustic setting away from permanent structures you have a little more latitude when it comes to design. This works out quite well with natural stone steps, because in many cases the natural sizes and shapes of the stones will inform the dimensions of the steps.

A stairway built using natural or cast cobbles, such as the left photo below, can be manipulated pretty easily to manage the rise in a fairly uniform way. However, it often makes sense from both a practical and an aesthetic point of view, to design the steps as a series of landings. Here, uniform landing depth is desirable, but it is not required and you may want to vary it a little bit so your step accommodates a slope more naturally.

A series of large, flat stones can make very striking steps, with each flat stone making up a single tread. In such situations you'll need to do more grading and modification of the slope terrain to fit the dimensions of the stones.

In either of the cases above, adding a handrail is probably not required by codes. But it is always a good idea and it will be much appreciated by your visitors.

Rock Garden

Rock gardens offer a good way to landscape difficult sites. Sloped areas or sites with sandy soil, for instance, are unfavorable for traditional lawns but are ideal for rock gardens. A rock garden is also a good choice if you're looking for an alternative to traditional ground covers and garden beds. Rock gardens traditionally feature hardy, alpine plant varieties that typically require infrequent watering.

Building a rock garden requires excavating the site (preferably a sloping or terraced area) and preparing the soil, placing the rocks, and planting. Moving and positioning large rocks is the most difficult task. If your rock garden site is large, consider hiring a landscape contractor to deliver and place the rocks for you.

Rock gardens will look more natural if they're built with stones that are all the same variety—or at least with stones that are similar in appearance. Using stone like that found in natural outcroppings in your area is a good idea. On the other hand, gardens with a larger variety of stone types have more visual variety and high potential for a very dramatic appearance when arranged with some skill.

Common Rock Garden Plants ▸

- Hens-and-Chicks
- Snow-in-Summer
- Coral bells
- Sedum
- Dianthus
- Rock jasmine
- Rockcress
- Dwarf juniper
- Yarrow

Tools & Materials ▸

- Wheelbarrow
- Shovel
- Work gloves
- Eye protection
- Mulch
- Eye protection and work gloves
- Course sand
- Pea gravel or rock chips
- Moss
- Buttermilk or yogurt
- Putty knife
- Alpine plants

A rock garden is an accessible first natural stone project. Rock gardens make a statement in a landscape and they are an attractive alternative to traditional garden beds.

How to Build a Rock Garden

The best site for a rock garden is sloping or terraced, ideally with southeast exposure. It should be completely free of deep-rooted weeds. If no such space is available, build up a raised bed with southeast exposure. The alpine plants traditionally used in rock gardens require soil with no clay and with excellent drainage. If the existing soil has clay, remove any ground cover and excavate the site to a depth of around 18". Replace the soil with equal parts loam, peat moss, and coarse sand.

Begin at the base of the site, placing the most substantial stones first. The idea is to create the impression of a natural subterranean rock formation weathered from exposure. Set the stones in the soil so they are at least half buried and so their most weathered surfaces are exposed. Slightly less than half of the surface of the site should be rock. Avoid even spacing or rows of rocks. Once the rocks are in place, cover the soil with a mulch of complementary pea gravel or rock chips.

"New" rocks moved to the site will need to weather before they will look natural. Encourage weathering by promoting moss and lichen growth. In a blender, combine a handful of moss with a cup of buttermilk or yogurt. Distribute the mixture on the exposed faces of rocks with a putty knife to promote moss growth.

After the rocks are in place for a few days and the soil has settled, begin planting the garden. As with the rocks, focus on native plant varieties and strive for natural placement, without excessive variety. You can plant several sizes, from small trees or shrubs to delicate alpine flowers. Place plants in crevices and niches between rocks, allowing them to cascade over the surface of the rock.

Stone Moon Window

You can build circular openings into brick or stone walls using a single semicircular wood form. Moon windows can be built to any dimension, although lifting and placing stones is more difficult as the project grows larger, while tapering stones to fit is a greater challenge as the circle gets smaller. To minimize the need for cutting and lifting stone, we built this window two feet in diameter atop an existing stone wall. Before doing this, you'll need to check with your local building inspector regarding restrictions on wall height, footings, and other design considerations. You may need to modify the dimensions to conform with the local building code.

Make sure to have at least one helper on hand. Building with stone is always physically demanding, and steps such as installing the brace and form (opposite page) require a helper.

Tools & Materials ▸

- Jigsaw
- Circular saw
- Drill
- Tape measure
- Level
- Chalk
- Mortar box
- Mason's hoe
- Trowels
- Jointing tool or tuck-pointer
- Mortar bag
- Stone chisel
- Mason's string
- Maul
- ¾" plywood
- ¼" plywood
- Wallboard screws (1 and 2")
- Tapered shims
- 2 × 4 and 2 × 8 lumber
- 4 × 4 posts
- Type M mortar (stiff mix)
- Ashlar stone
- Eye protection and work gloves

A moon window is just about the most dramatic garden element you can build. We constructed the wall shown here using cut ashlar mortared around a semicircular form, but using brick is also an option. Once the bottom half of the window has set up, the form is flipped and the top stones are placed. The construction technique for the form is the same one used in building an arch (pages 156 to 159).

How to Build a Stone Moon Window

Build a plywood form, following the instructions on page 157. Select stones for the top of the circle with sides that are squared off or slightly tapered. Dry-lay the stones around the outside of the form, spacing the stones with shims that are roughly ¼" thick at their narrow end.

Number each stone and a corresponding point on the form using chalk, and then set the stones aside. Turn the form around, and label a second set of stones for the bottom of the circle. *Tip: To avoid confusion, use letters to label the bottom set of stones instead of numbers.*

Prepare a stiff mix of Type M mortar and lay a ½"-thick mortar bed on top of the wall for the base of the circle. Center the stone that will be at the base of the circle in the mortar.

Set the form on top of the stone and brace the form by constructing a sturdy 2 × 4 scaffold and secure it by constructing a bracing structure made from 4 × 4 posts and 2 × 4 lumber. We used pairs of 2 × 4s nailed together for lengthwise supports. Check the form for level in both directions and adjust the braces as required. Screw the braces to the form so the edges are at least ¼" in from the edges of the form.

(continued)

Extend the mortar bed along the wall and add stones, buttering one end of each stone and tapping them into place with a trowel. Keep the joint width consistent with the existing wall, but set the depth of new joints at about 1" to allow for tuck-pointing.

Attach mason's string at the center of the front and back of the form and use the strings to check the alignment of each stone.

Stagger the joints as you build upward and outward. Alternate large and small stones for maximum strength and a natural look. Stop occasionally to smooth joints that have hardened enough to resist minimal finger pressure.

Dress stones as necessary to follow large bulges or curves as you lay stones around the circle. The sides should be roughly squared off.

Once you've laid stones about ½" beyond the top edge of the form, disassemble the bracing.

Invert the form on top of the wall in preparation for laying the top half of the circle. The bottom edge of the form should be set roughly ½" higher than the top of the lower half of the circle. Check the braces for level (both lengthwise and widthwise), and adjust them as necessary and reattach them to the posts.

Lay stones around the circle, working from the bottom up, so the top, or keystone, is laid last. If mortar oozes from the joints, insert temporary shims. Remove the shims after 2 hours and pack the voids with mortar.

Once the keystone is in place, smooth the remaining joints. Remove the form. Let the wall set up overnight, and then mist it several times a day for a week.

Remove any excess mortar from the joints inside the circle. Mist lightly, and then tuck-point all joints with stiff mortar so they are of equal depth.

A putty-like consistency will develop in joints. At this point, use a joining tool to smooth the surface. Let the mortar harden overnight. Mist the wall for 5 more days.

These convenient interlocking pavers are made with DIYers in mind. They are easy to install and often come with fully plotted patterns for simple design preparation and installation.

Cobblestone Paver Patio

Not technically natural stone, interlocking stone pavers have advanced by leaps and bounds from the monochromatic, cookie-cutter bricks and slabs associated with first-generation concrete pavers. The latest products feature subtle color blends that lend themselves well to organic, irregular patterns. A tumbling process during manufacturing can further age the pavers so they look more like natural cobblestones. The technological advances in the casting and finishing processes have become so sophisticated that a well-selected concrete paver patio could look as comfortable in a traditional Tuscan village as in a suburban backyard.

When choosing pavers for a patio, pick a style and blend of shapes and sizes that complement your landscape. Use your house and other stone or masonry in the landscape to inform your decisions on colors and shade. Be aware that some paver styles require set purchase amounts, and it's not always possible to return partly used pallets.

Here we lay a cobble patio that uses three sizes of stone. These may be purchased by the band (a fraction of a pallet), minimizing leftovers. Notice that an edge course creates a pleasing border around our patio. Bring a drawing of your patio with exact measurements to your stone yard. Based on your layout pattern, the sales staff will be able to tell you how much of each size stone you'll need to purchase.

One great advantage to interlocking concrete pavers is that they create a very rigid surface with high resistance to movement and sinking, even when set on a gravel base. This makes them suitable for driveways and busy walkways as well as backyard patios. If you prefer, you can set pavers into a mortar bed on a concrete slab.

Note: The differences do not bear on the installation process, but its helpful to distinguish between brick pavers and concrete pavers. Brick pavers are made of fired clay. Concrete pavers are cast from concrete that's placed in forms and cured. Natural cobblestones are small stones with flat, smooth surfaces.

Tools & Materials ▸

- Wheelbarrow
- Garden rake
- 4-ft. level
- Hand maul
- Small pry bar
- Wood stakes
- Chalk line
- Mason's string
- Line level
- Shovel
- Tape measure
- Square-nose spade
- 1"-dia. metal pipes
- Straight 2 × 4
- 4 × 4 squares of plywood
- Particle mask
- Water-cooled masonry saw
- Plate compactor
- Gloves, ear protection, and safety glasses
- Stiff bristle broom
- 6 × 6 cobble squares
- 6 × 9 cobble rectangles
- 3 × 6 cobble rectangles for edges
- Compactable gravel
- Coarse sand
- Paver edging and spikes
- Joint sand
- Flathead screwdriver

Cobblestones ▸

Today, the word "cobblestone" more often refers to cast concrete masonry units that mimic the look of natural cobblestones. Although they are tumbled to give them a slightly aged appearance, cast concrete cobbles are more uniform in shape, size, and color. This is an advantage when it comes to installation, but purists might object to the appearance.

Pattern Detail: Cobblestone Patio

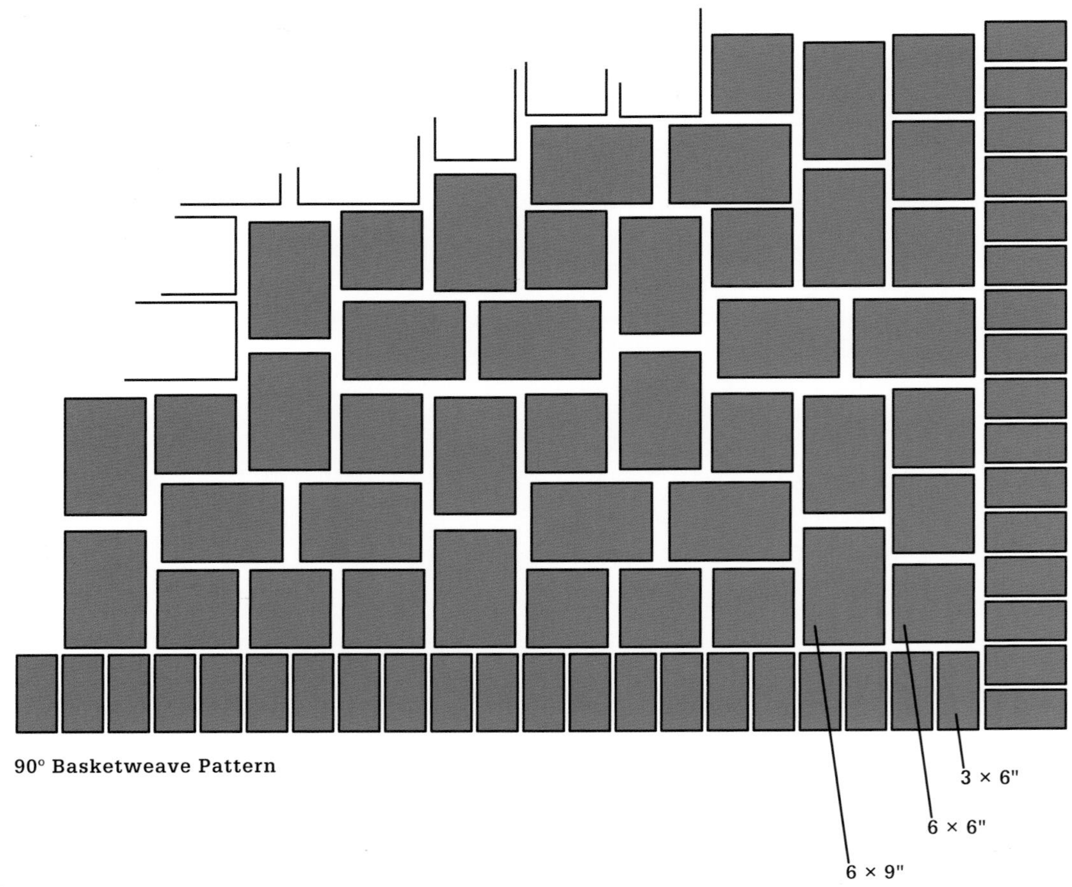

Cross-Section: Cobblestone Patio

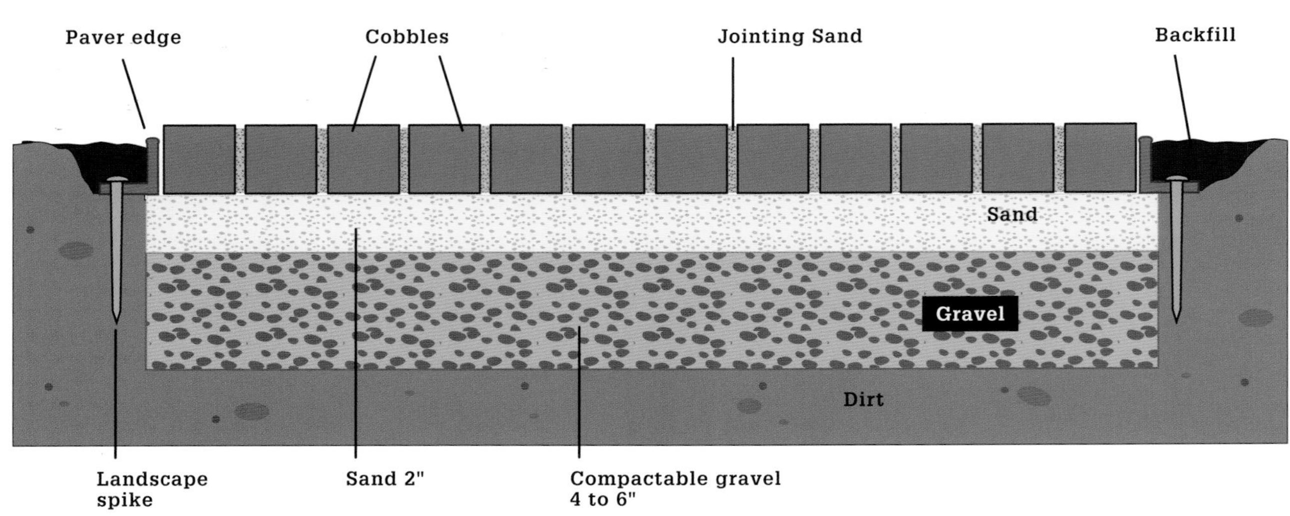

How to Build a Cobblestone Paver Patio

Lay out the patio outline and set digging depth with grade stakes. Factor a minimum 1" per 8 ft. of drainage slope away from the house if the patio is next to your home. Remove the strings.

Excavate the building site, paying close attention to the outlines and stakes that denote the excavation depth and slope.

Rake and screed the compactable gravel to follow the drainage slope and then compact with a plate compactor.

Lay out square corners for the patio with stakes and string, starting next to an adjoining structure. Use the 3-4-5 method to check the intersecting lines for squareness at the corners.

(continued)

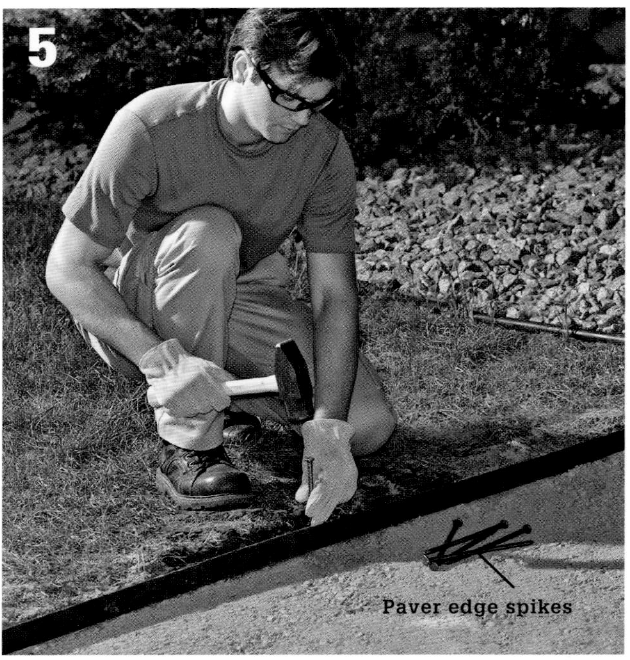

Snap chalk lines directly below the outlines you've created with the mason's strings and then install professional-grade paver edging at the lines. The paver edge should rest on compacted gravel, not soil.

Strike off the coarse sand base by dragging a 2 × 4 screed across 1" pipes that are used as screed gauges.

Begin laying out the cobbles, starting at the square corner. Work in small sections on approximately 5 square ft. Simply set the cobbles into the sand base—do not adjust them from side to side or try to reset the height.

Tie additional strings to establish a guide line that bisects the project and is perpendicular to the bond line at the end of the layout pattern.

Creating a Layout ▸

The number of purchasing options available when you shop for pavers makes it possible to create just about any patio layout pattern you can imagine. There is nothing stopping you from going wild and creating a layout that's truly creative. However, most landscape centers are happy to work with you to create a layout that employs tested design ideas and consumes pavers in a very efficient manner with as little cutting as possible.

Another option for DIY designers is to visit the website of the paver manufacturer (you should be able to get the information from your paver dealer). Many of these have applications where you can choose a basic style you like (such as the basketweave pattern seen here), and enter size information. You'll receive a printout of what the pattern should look like along with a shopping list for the materials you'll need, all the way down to sand and spikes for your paver edging. To see an example of a design calculator/estimator, visit the website for Borgert Products, maker of the cobbles seen here in a 30% square, 70% rectangle basketweave pattern with rectangular border.

Install the paver edging for the rest of the patio using the bond lines as reference. Brush sand out of the edging installation area so the paver edging rests on the compactable base. Replace and smooth the sand after the edging is installed.

Cut the cobbles to fit the layout using a wet saw (rentable).

(continued)

Fill gaps between cobbles with decorative sand, such as crushed granite, or with specially formulated jointing sand that hardens when dampened for a more formal look.

Tamp the cobble stones with a plate compactor to bring them to level and seat them in the base. Add jointing sand in the joints if levels drop as you work.

Making Curves ▸

At rounded corners and curves, install border pavers (below left) in a fan pattern with even gaps between the pavers. Gentle curves may accommodate full-size border pavers, but for sharper turns you usually need to cut tapers into the paver edges so you don't end up with wide gaps at the outside. When using border pavers in a curved layout, the field pavers will need to be trimmed to fit the odd spaces created where the field and borders intersect (below right).

Variation: Circular Paver Patio

Set the first ring of pavers around the center paver. Check their positions carefully, and make sure the spacing lugs are oriented correctly. If the pavers don't have lugs, gap them according to the manufacturer's specifications. *Note: Do not hammer or tamp the pavers into the sand bed unless the manufacturer directs otherwise.*

Set the remaining pavers, completing each ring according to your layout diagram. Be sure to offset the paver joints between rows. The pavers may be labeled, requiring them to be installed in a specific order as you work around the circle. After a sizable area is laid, work from your plywood platform set atop the pavers.

Install rigid paver edging along the patio's perimeter. Set the edging on top of the gravel sub-base but not the sand bed. *Tip: Dampening the sand bed along the patio edge makes it easy to cut the sand away cleanly with a trowel before setting the edging.*

Inspect the paving to make sure all joints are aligned properly and all gaps are consistent. Make minor adjustments to pavers as needed using a flathead screwdriver as a pry bar. Be careful not to mar the paver edges as you pry.

Sandset flagstone patios blend nicely with natural landscapes. Although flagstone evokes a natural feel, the patio can appear rustic or formal. This patio has clean, well-tamped joints and straight, groomed edges along the perimeter that lend a formal feel. Plantings in the joints or a rough, natural perimeter would give the same patio a more relaxed, rustic feel.

Sandset Flagstone Patio

Flagstones make a great, long-lasting patio surface with a naturally rough texture and a perfectly imperfect look and finish. Randomly shaped stones are especially suited to patios with curved borders, but they can also be cut to form straight lines. Your patio will appear more at home in your landscape if the flagstones you choose are of the same stone species as other stones in the area. For example, if your gravel paths and walls are made from a local buff limestone, look for the same material in limestone flags.

Flagstones usually come in large slabs, sold as flagstone, or in smaller pieces (typically 16" or smaller), sold as *steppers*. You can make a patio out of either. Larger stones will make a solid patio with a more even surface, but the bigger ones can require three strong people to position, and large stones are hard to cut and fit tightly. If your soil drains well and is stable, flagstones can be laid on nothing more than a layer of sand. However, if you have unstable clay soil that becomes soft when wet, start with a four-inch-thick foundation of compactable gravel under your sand.

There are a few different options for filling the spaces between flagstones. One popular treatment is to plant them with low-growing perennials suited to crevice culture. For best results, use sand-based soil between flagstones when planting. Also, stick to very small plants that can withstand foot traffic. Another option is to create a rock garden effect by eliminating an occasional small flag in an out-of-the-way spot and planting the space with a sturdy accent species. If you prefer not to have a planted patio, simply fill the joints with sand or fine gravel—just be sure to add landscape fabric under your sand base to discourage weed growth.

The following project includes steps for building a classic flagstone patio. You'll also find instructions for building low dry stone walls, the ultimate add-on to a stone patio surface. If you're new to working with natural stone, see pages 177 to 179 for some basic cutting tips.

Tools & Materials ▸

Mason's string
Line level
Rope or hose
Excavation tools
Spud bar
Broom
Stakes
Marking paint
1" (outside diameter) pipe
Coarse sand
Straight 2 × 4
Flagstone
Spray bottle
Stone edging
Sand-based soil or joint sand
Lumber (2 × 2, 2 × 4)
Drill
Mason's trowel
Stiff-bristle brush
Circular saw with masonry blade
Plugs or seeds for ground cover
Eye and ear protection
Work gloves
¾" plywood
3½" deck screws
Pointing chisel
Pitching chisel
Stone chisel
Hand maul
Dust mask
Chalk or a crayon
Square-nose spade
Crushed stone
Ashlar
Mortar
Capstones

Adding a Stone Wall ▸

A dry stone wall is a simple, beautiful addition to a flagstone patio. A wall functions as extra seating, a place to set plants, or extra countertop or tabletop space. It also provides visual definition to your outdoor space. See page 182 for how to build a stone wall.

Construction Details: Sandset Flagstone Patio

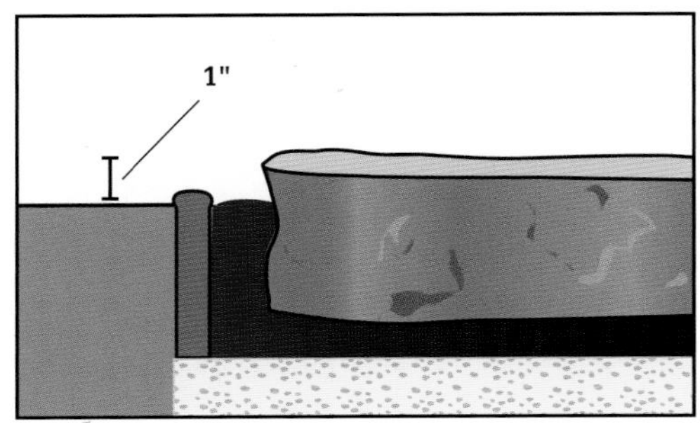

Lay flagstones so their tops are approximately ½ to 1" above the surrounding ground. Because natural stones are not uniform in thickness, you will need to adjust sand or dirt beneath each flagstone, as needed.

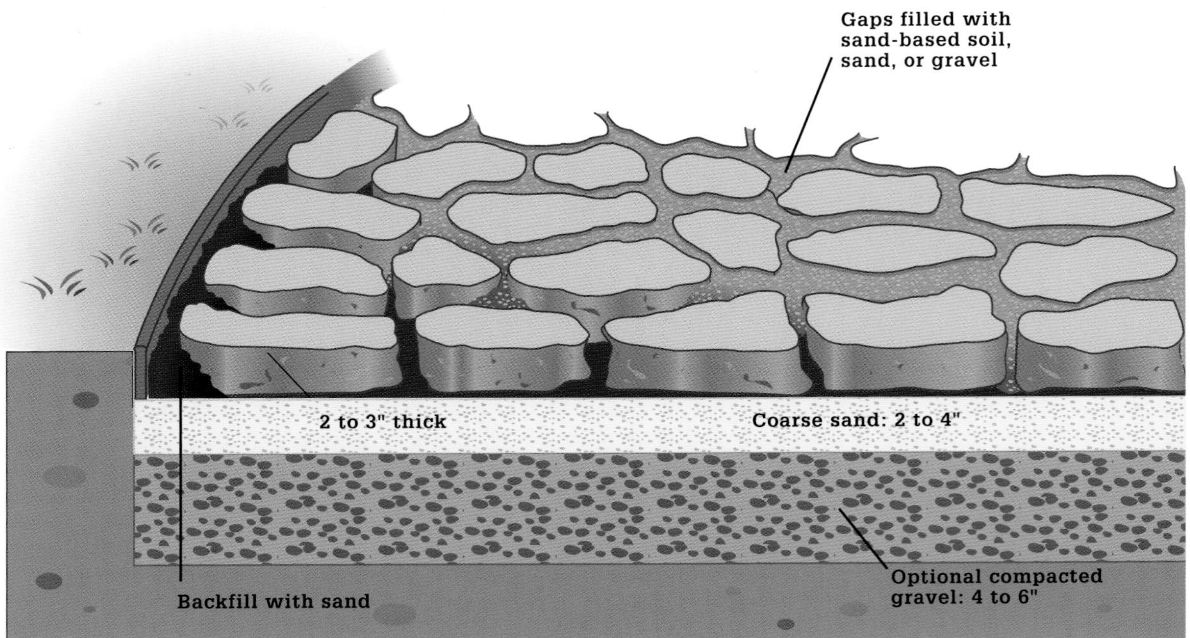

A typical sandset patio has a layer of coarse sand for embedding the flagstones. A sub-base of compactable gravel is an option for improved stability and drainage. The joints between stones can be filled with sand, gravel, or soil and plants. Edging material is optional.

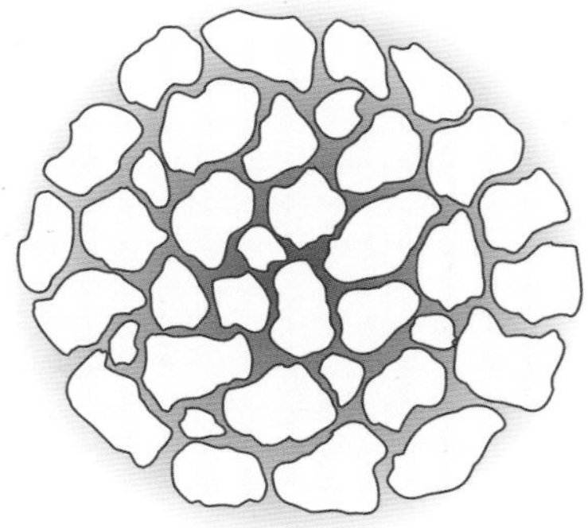

Irregular flagstones look natural and are easy to work with in round layouts.

Flagstones that are cut into rectangular shapes can be laid in square or rectangular patterns with uniform gaps.

How to Build a Sandset Flagstone Patio

Outline the patio base using string and stakes for straight lines and/or a rope or hose for curves. The base should extend at least 2 to 4" beyond the edges of the flagstones, except where the patio will butt up to a wall. Transfer the outline to the ground with marking paint. Remove any sod and vegetation within the base area.

Set up layout strings to guide the excavation using stakes or batterboards (see pages 38 to 41 for detailed steps on layout and site preparation). Excavate the base to a depth of 2" plus the stone thickness plus ½ to 1". Slope the ground away from the house foundation at a rate of ¼" per ft.

Lay sections of 1" pipe across the project area to serve as screed gauges. These allow you to strike off sand at a consistent depth when you drag a screed board over them. *Note: Since large flagstones can be held in place adequately by the surrounding soil, edging for the patio is optional; it often looks best to allow neighboring ground cover to grow up to the edges of the stones. If you do plan to use edging, install it now.*

Fill the site with coarse sand slightly above the screed gauges. With a helper, drag a straight 2 × 4 across the screed gauges to level off the sand. Use a screed board that's long enough so that you can avoid stepping in the sand. Work the screed in a back-and-forth sawing motion. Remove the pipes once each section is finished, fill in the voids, and smooth the surface flat.

(continued)

Arrange your flagstones into groups according to size and shape. As a general rule, start paving with the broadest stones and fill in around them with increasingly smaller pieces, but appearance and sight lines are also important: If there is one nice stone with a flat surface and good color, feature it in the center of the patio. Or, if some of the patio will be visible from the house, choose nicer stones for these areas.

Begin by laying large, thick stones around the perimeter of the patio. Leave a consistent gap of about 1" between stones by matching pieces like a puzzle and cutting and dressing stones as needed (see pages 177 to 179). The outer edge of the patio should form smooth curves (or straight lines) without jutting pieces or abrupt irregularities. Level stones as needed by prying up with a spud bar and adding or removing sand underneath.

Fill in around the larger stones with smaller pieces cut to fit the spaces, as needed, working from the outside in. After setting a band of stones a few courses wide, lay a 2 × 4 across the stones to make sure they're level with one another. Add or remove sand below to adjust their height and dampen the sand occasionally to make it easier to work with.

Fill the joints between stones with sand-based, weed-seed-free soil (see page 231). Sweep the soil across the patio surface to fill the cracks, and then water the soil so it settles. Repeat as needed until the soil reaches the desired level. Plant plugs or seeds for ground cover to grow up between the stones, if desired.

Variation: To finish the patio with sand instead of soil and plants, spread sand over the patio and sweep across the stones with a push broom to fill the joints. Pack the sand with your fingers or a piece of wood. Spray the entire area with water to help compact the sand. Let the patio dry. Repeat filling and spraying until the joints are full and the stones are securely locked in place.

Choosing Soil & Plants for Your Patio ▸

Sand-based soil (also called "patio planting" soil) is the best material to use for planting between flagstones. This mixture of soil and sand sweeps easily into joints, and it resists tight compaction to promote healthy plant growth, as well as surface drainage. Regular soil can become too compacted for effective planting and drainage and soil from your yard will undoubtedly contain weeds. Sand-based soil is available in bulk or by the bag and is often custom-mixed at most large garden centers.

As for the best plants to use, listed below are a few species that tend to do well in a patio application. Ask a local supplier what works best for your climate.

- Alyssum
- Rock cress
- Thrift
- Miniature dianthus
- Candytuft
- Lobelia
- Forget-me-not
- Saxifrage
- Sedum
- Thymus
- Scotch moss
- Irish moss
- Woolly thyme
- Mock strawberry

Patio "planting soil" (for planting between stones) is available in bulk or bags at most garden centers. It is good for filling cracks because the sand base makes it dry and smooth enough to sweep into cracks, yet the black compost will support plant growth. Because it is bagged, you can be assured it doesn't come with weeds.

Pebbled Stepping Stone Path

Stepping stones in a path do two jobs: they lead the eye and they carry the traveler. In both cases the goal is rarely fast direct transport, but more of a relaxing meander that's comfortable, slow-paced, and above all natural. Arrange the stepping stones in a walking path according to the gaits and strides of the people most likely to use the pathway. Realize that our gaits tend to be longer on a utility path than in a rock garden.

Sometimes steppers are placed more for visual effect, with the knowledge that they will break the pacing rule with artful clusters of stones. Clustering is also an effective way to slow or congregate walkers near a Y in the path or at a good vantage point for a striking feature of the garden.

Choose steppers and pebbles that are complementary in color. Shades of medium to dark gray are a popular combination for a Zen feeling. Too much contrast or very bright colors tend to undermine the sense of tranquility a pebbled stepping stone path can achieve.

Tools & Materials ▸

- 1 × 3 stakes
- Lumber (1 × 2, 2 × 4)
- Mason's string
- Mallet
- Hose or rope
- Landscaping paint
- Measuring tape
- Edging
- Spun-bonded landscape fabric
- Sod stripper
- Coarse sand
- Thick steppers or broad river rocks with one flat face
- ¼ to ½" pond pebbles
- 2½"-dia. river rock
- Wheelbarrow
- Round-nosed spade
- Edging/trenching spade
- Flat-nosed spade
- Hand tamper
- Work gloves

Stepping stones blend beautifully into many types of landscaping, including rock gardens, ponds, flower or vegetable gardens, or manicured grass lawns.

How to Build a Pebbled Stepping Stone Path

Excavate the pathway site and prepare a foundation. Substitute coarse building sand for compactable gravel. Strike the sand to a consistent depth with a notched 2 × 4.

Level the stones by adding and removing sand until they are solidly seated. On flat runs, you should be able to rest a flat 2 × 4 on three stones at once, making solid contact with each. It is much easier to pack sand under stones if you moisten the sand first. Also moisten the sand bed to prevent sand from drifting.

Spread out a layer of the largest diameter rock if you are using two or more series of infill.

Add the smallest size infill stones last, spreading them evenly so you do not have to rake them much.

Place the remaining larger-diameter infill stones around the surface of the walkway to enhance the visual effect of the pathway.

Zen Garden

What's commonly called a Zen garden in the West is actually a Japanese dry garden, with little historical connection to Zen Buddhism. The form typically consists of sparse, carefully positioned stones in a meticulously raked bed of coarse sand or fine gravel. Japanese dry gardens can be immensely satisfying. Proponents find the uncluttered space calming and the act of raking out waterlike ripples in the gravel soothing and perhaps even healing. The fact that they are low maintenance and drought resistant is another advantage.

Site your garden on flat or barely sloped ground away from messy trees and shrubs (and cats), as gravel and sand are eventually spoiled by the accumulation of organic matter. There are many materials you can use as the rakable medium for the garden. Generally, lighter-colored, very coarse sand is preferred—it needs to be small enough to be raked into rills yet large enough that the rake lines don't settle out immediately. Crushed granite is a viable medium. Another option that is used occasionally is turkey grit, a fine gravel available from farm supply outlets. In this project, we show you how to edge your garden with cast pavers set on edge, although you may prefer to use natural stone blocks or even smooth stones in the four to six inches range.

Tools & Materials ▸

- Stakes
- Mason's string
- Garden hose
- Landscape marking paint
- Straight 2 × 4
- Level
- Measuring tape
- Compactable gravel
- Crushed granite (light colored)
- Excavating tools
- Hand maul
- Manual tamper
- Spun-bonded landscape fabric
- Fieldstone steppers
- Specimen stones
- Border stones or blocks
- Eye protection and work gloves

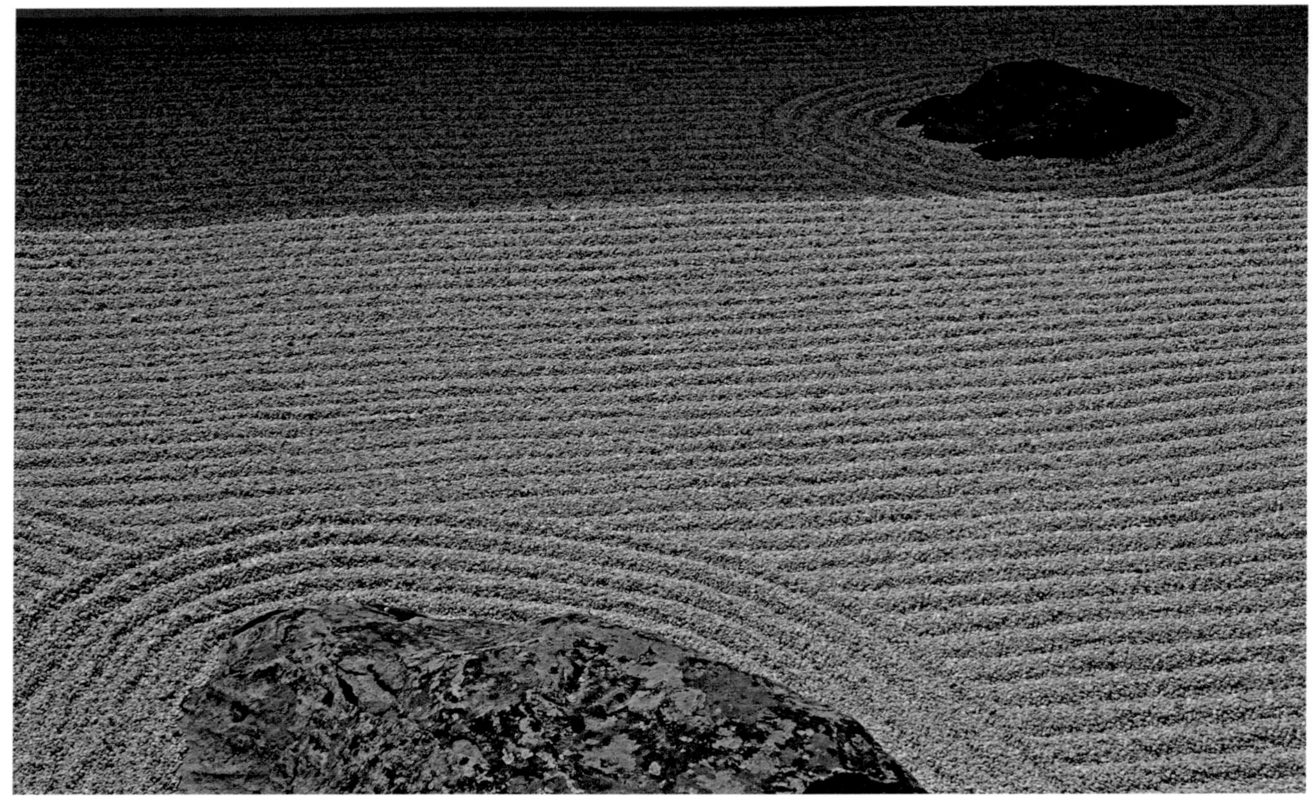

A Zen garden is a small rock garden, typically featuring a few large specimen stones inset into a bed of gravel. It gets its name from the meditative benefits of raking the gravel.

How to Make a Zen Garden

Lay out the garden location using stakes and string or hoses and then mark the outline directly onto the ground with landscape paint.

Excavate the site and install any large specimen stones that require burial more than ½ ft. below grade.

Dig a trench around the border for the border stones.

Pour a 3" thick layer of compactable gravel into the border trench and tamp down with a post or a hand tamper.

(continued)

Place border blocks into the trench and adjust them so the tops are even.

Test different configurations of rocks in the garden to find an arrangement you like. If it's a larger garden, strategically place a few flat rocks so you can reach the entire garden with a rake without stepping in the raking medium.

Set the stones in position on individual beds of sand about 1" thick.

Rake the medium into pleasing patterns with a special rake (see next page).

How to Make a Zen Garden Rake ▸

Once you have constructed your Zen garden, you will use two tools to interact with it: your eyes and a good rake. While any garden rake will suffice for creating the swirling and concentric rills that are hallmarks of the Zen garden, a special rake that's dedicated to the garden will enhance your hands-on interaction.

Many Zen garden rakes are constructed from bamboo. Bamboo is lightweight and readily available, especially through internet sites. While you can certainly choose this material, you're likely to find that the lightness can actually work against it, causing you to exert more strain to cut through the raking medium. A rake made from solid wood has greater heft that lets it glide more smoothly through the medium. The rake shown here is made using only the following materials:

- 1¼"-dia. by 48" oak or pine dowel (handle)
- ½" by 36" oak or pine dowel (tines)
- 2 × 3 × 9½" piece of red oak (head)

Figure 1

Figure 2

Figure 3

Here is how to make it. Start by sanding all of the stock very smooth using sandpaper up to 150 grit in coarseness. Soften the edges of the 2 × 3 with the sandpaper. Drill a 1¼" dia. hole in the head for the handle (Figure 1). The hole should go all the way through the head at a 22½° downward angle (half of a 45° angle), with the top of the hole no closer than ¾" to the top of the head. Use a backer board when drilling to prevent blowout and splinters.

Next, drill ½"-dia. by 1"-deep seat holes for the tines in the bottom edge of the blank. Locate centers of the two end holes 1" from the ends. Measure in 2½" from each end hole and mark centers for the intermediate tines. Use masking tape to mark a drilling depth of 1" on your drill bit and then drill perpendicular holes at each centerline.

Cut four 5"-long pieces of the ½"-dia. oak doweling for the tines. Apply wood glue into the bottom of each hole and insert the tines, setting them by gently tapping with a wood mallet (Figure 2). Then, apply glue to the handle holes sides and insert the handle so the end protrudes all the way through. After the glue dries, drill a ½"-dia. hole down through the top of the head and into the handle. Glue a ½" dowel into the hole to reinforce the handle (this is called pinning).

Finally, use a back saw, gentleman's saw, or Japanese flush-cutting saw to trim the handle end and the handle pin flush with the head (Figure 3). Sand to smooth the trimmed ends and remove any dried glue. Finish with two or three light coats of wipe-on polyurethane tinted for red oak.

Decorative Masonry Finishing

Add textural interest to masonry surfaces with stains, stucco, veneer, and other decorative applications that dress up a project. You'll learn skills for dressing up basic concrete and how to use stone as the icing on your project. The techniques in this chapter can be applied to many of the projects in this book.

In this chapter:

Stamped Concrete Finishes

Stamped finishes can bring interesting texture to ordinary concrete sidewalks, patios, and driveways. Stamping mats are available in a variety of textures and patterns, and they can be rented at most equipment rental centers and concrete supply stores.

As you plan your concrete project, also plan the layout of the stamping mats to help maintain a consistent pattern across the project. For best results, mark a reference line at or near the center of the project and align the first mat with it. Align the subsequent mats with the first, working outward toward the ends of the project. Plan for long seams to fall across the project rather than along the length of it to avoid misaligned seams. You may have to hand-finish textures at corners, along sides, or near other obstructions using specialty stamps or an aluminum chisel.

The stamping pads should be pressed into slightly stiff concrete to a depth of about one inch. Professionals typically use enough pads to cover the entire project area. For DIYers, it probably makes more sense to have one or two pads and reuse them.

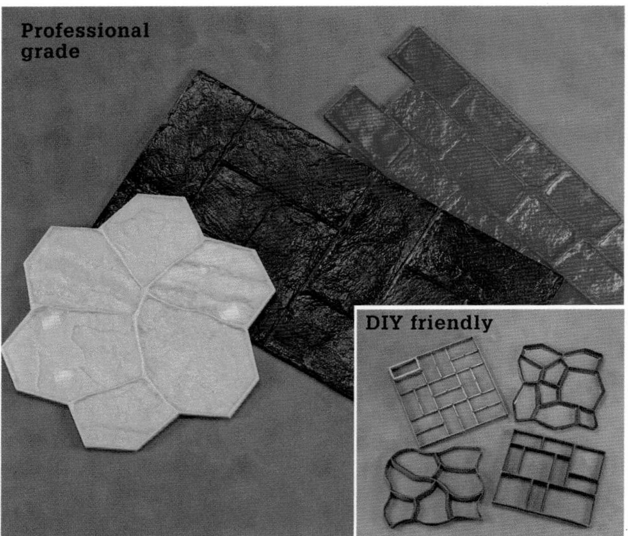

Stamping mats can be pressed onto fresh concrete, as with the professional-grade mats in the larger photo above (these are fairly expensive). DIY-oriented stamping mats are cheaper but offer fewer options. The DIY mats are usually open grids, so they can be used as either stamp pads or molds.

Tools & Materials ▸

Tools for mixing and pouring concrete
Hand tamp
Textured stamping mats
Aluminum chisel
Materials for mixing, pouring, and coloring concrete
Pressure washer
Powder release agent
Work gloves

Stamped concrete can emulate the appearance of expensive imported stones or just about any other paver, but at a much lower cost.

How to Stamp a Concrete Surface

Pour a concrete slab and mark a reference line for the first mat at or near the center of the form. When the bleed water disappears from the surface, toss powder release agent across the surface in the amount specified in the manufacturer's instructions.

Align the first stamping mat with the reference line on the form, following your layout plan. Once the mat is placed, do not adjust it. Carefully step onto the mat or use a hand tamp to embed the stamp into the concrete.

Butt the second pad against the first, so the seams are flush and aligned. Embed the mat into the concrete, and then place a third mat, maintaining the continuous pattern. Remove and reuse mats. When the project area is wider than the stamping pads, complete rows across the width before stamping lengthwise.

After the concrete has cured for three days, remove leftover release agent from the surface using a pressure washer with a wide-fan spray—work in smooth, even strokes no more than 24" from the surface. Allow the concrete to cure fully for an additional week, and then apply an acrylic concrete sealer according the manufacturer's instructions. Remove the forms and backfill.

Acid-stained Concrete Patio

Acid staining is a permanent color treatment for cured concrete that yields a translucent, attractively mottled finish ideally suited to patios. Unlike paint or pigmented concrete stains, both of which are surface coatings, acid stain is a chemical solution that soaks into the concrete pores and reacts with the minerals to create the desired color. The color doesn't peel or flake off, and it fades very little over many years. Acid stain won't hide blemishes or discoloration in the original concrete surface, but many consider this an important part of its natural appeal. If your patio or walkway is fully exposed, bear in mind that some colors of acid stain may fade in direct sunlight, so be sure to choose a color guaranteed by the manufacturer not to fade.

You can apply acid stain to new concrete that has cured for at least four to six weeks (check the stain manufacturer's requirements for curing times) or old concrete that is free of any previously applied sealants. Test old concrete by spraying the surface with water: If the water beads on the surface instead of soaking in, there's probably a sealer on there, and it must be removed for good results with the stain. Ask the stain manufacturer for recommended concrete sealer remover products to use.

Another important preparation step with either new or old concrete is to color-test a few shades of stain on the concrete you'll be working with. Stain suppliers often sell sample-size quantities of stain for this purpose. Since acid stain affects every surface a little differently, it's worth the effort to run a test before committing to a color. Be sure to test in an inconspicuous area, because the stain can't be removed once it's applied.

Tools & Materials ▸

- Tape
- Plastic sheeting
- Stiff-bristled brush
- Concrete cleaner and solvent
- All-plastic garden sprayer
- Stain
- Protective clothing
- Eye protection
- NIOSH/MSHA-approved respirator
- Cleaner/neutralizer
- Medium-bristled scrub brush
- Towel
- Sealer

Acid-staining a patio is a fairly simple procedure that can be done by almost anyone. Be sure the surface is completely cured, clean (use water, not soap), and dry; and be sure to wear the protective gear, as recommended by the manufacturer. A properly maintained acid-stained patio can last forever.

How to Apply an Acid Stain

Protect all surfaces adjacent to the concrete and any nearby plants with tape and plastic sheeting.

Clean the entire surface with an approved cleaner and stiff-bristled brush. Mix the stain with water in an all-plastic garden sprayer, as directed. Use an approved solvent to remove undesirable markings on the concrete surface (stain won't hide them). Rinse thoroughly and then let dry.

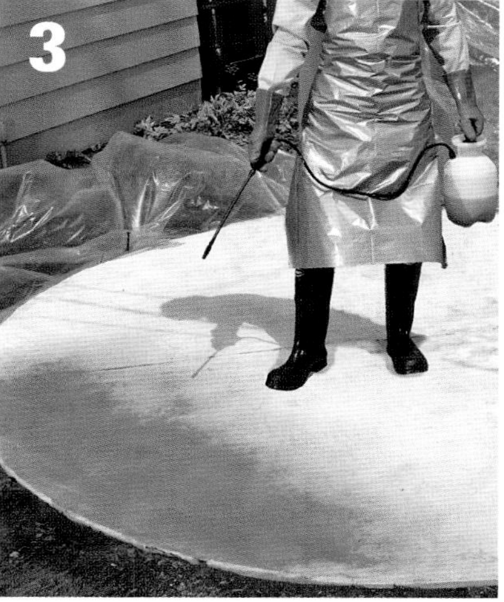

Spray the stain onto the concrete using random circular motions and holding the spray tip about 18" from the surface. Work backward from one side. Maintain pressure on the sprayer so the spray pattern is consistently fine and even. Wet the surface completely, but avoid creating puddles. Allow the first coat of stain to dry completely.

Warning ▸

Always pour acid stain into water; never pour water into acid stain.

Apply a second coat using the same technique. Darker tones will appear with the second coat; the wetter the surface, the darker the tones will be (but again, avoid puddles of stain). Let the second coat dry completely.

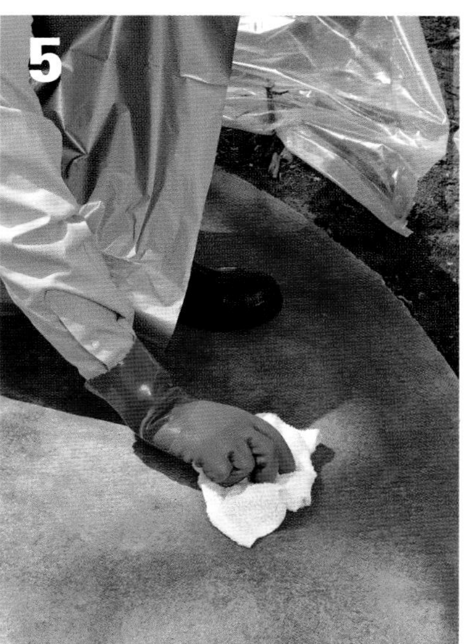

Wash the surface using a recommended cleaner/neutralizer and a medium-bristled scrub brush to remove dried stain residue. Thoroughly rinse according to the manufacturer's instructions. Test-wipe the surface with a white towel: If the towel shows stain, wash and rinse the surface again.

Apply sealer using a high-quality deck/patio sprayer (or other recommended applicator). Follow the manufacturer's instructions and recommendations—most sealers require multiple coats. Maintain even pressure on the sprayer for a consistently fine spray. Wet the surface completely with a thin coating, and avoid puddles. Let the sealer dry between coats.

Decorative Concrete Floor

Most people are accustomed to thinking of concrete primarily as a utilitarian substance, but it can also mimic a variety of flooring types and be a colorful and beautiful addition to your basement room.

Concrete is a hard and durable building material, but it is also porous—so it is susceptible to staining. Many stains can be removed with the proper cleaner, but sealing and painting prevents oil, grease, and other stains from penetrating the surface in the first place; and cleanup is a whole lot easier.

Even after degreasing a concrete floor, residual grease or oils can create serious adhesion problems for coatings of sealant or paint. To determine whether your floor has been adequately cleaned, pour a glass of water on the concrete floor. If it is ready for sealing, the water will soak into the surface quickly and evenly. If the water beads, you may have to clean it again. Detergent used in combination with a steam cleaner can remove stubborn stains better than a cleaner alone.

There are four important reasons to seal your concrete floor: to protect the floor from dirt, oil, grease, chemicals, and stains; to dust-proof the surface; to protect the floor from abrasion and sunlight exposure; and to repel water and protect the floor from freeze-thaw damage.

Tools & Materials ▸

Concrete cleaner
Acid-tolerant pump sprayer
Alkaline-base neutralizer
Sealant
Rubber boots
Rubber gloves
Roller tray
Wet vacuum
Acid-tolerant bucket
Eye protection and work gloves
Garden hose with nozzle
Paint roller and tray
Primer
Painter's tape
Plastic sheeting
Garden sprayer
Concrete stain
Soft-woven roller cover
High-pressure washer
Paintbrush
Respirator
Stiff-bristle broom and brush
Extension handle

Etching and sealing a concrete floor that is in good condition yields a slick-looking surface that has a contemporary feel and is easy to maintain.

How to Seal Concrete Basement Floors

Clean and prepare the surface by first sweeping up all debris. Next, remove all surface muck: mud, wax, and grease. Finally, remove existing paints or coatings.

Saturate the surface with clean water. The surface needs to be wet before acid etching. Use this opportunity to check for any areas where water beads up. If water beads on the surface, contaminants still need to be cleaned off with a suitable cleaner or chemical stripper.

Test your acid-tolerant pump sprayer with water to make sure it releases a wide, even mist. Once you have the spray nozzle set, check the manufacturer's instructions for the etching solution and fill the pump sprayer (or sprinkling can) with the recommended amount of water.

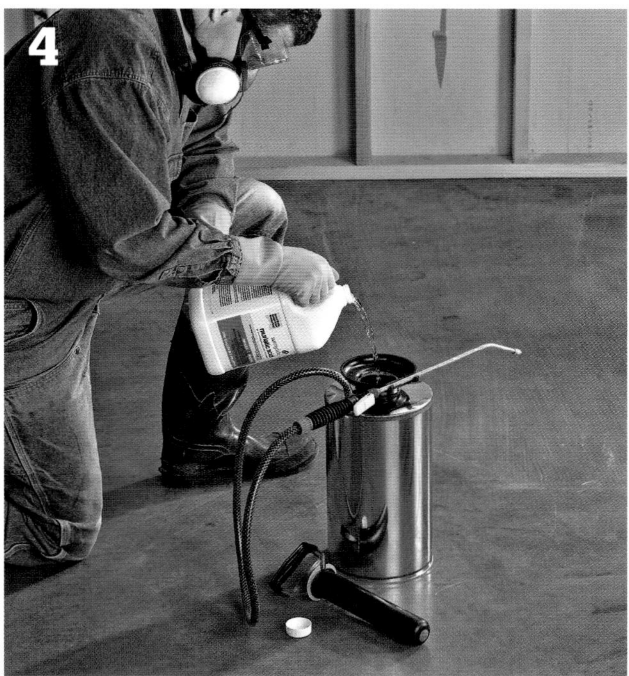

Add the acid-etching contents to the water in the acid-tolerant pump sprayer. Follow the directions (and mixing proportions) specified by the manufacturer. Use caution and wear safety equipment.

(continued)

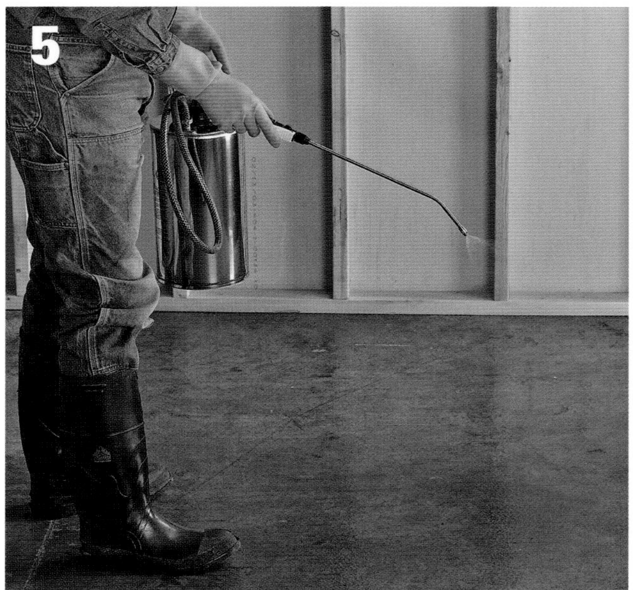

Apply the acid solution. Using the sprinkling can or acid-tolerant pump spray unit, evenly apply the diluted acid solution over the concrete floor. Do not allow acid solution to dry at any time during the etching and cleaning process. Etch small areas at a time, 10 × 10 ft. or smaller. If there is a slope, begin on the low side of the slope and work upward.

Use a stiff-bristle broom or scrubber to work the acid solution into the concrete. Let the acid sit for 5 to 10 minutes, or as indicated by the manufacturer's directions. A mild foaming action indicates that the product is working. If no bubbling or fizzing occurs, it means there is still grease, oil, or a concrete treatment on the surface that is interfering. If this occurs, follow steps 7 to 12 and then clean again.

Once the fizzing has stopped, the acid has finished reacting with the alkaline concrete surface and formed pH-neutral salts. Neutralize any remaining acid with an alkaline-base solution. Put 1 gal. of water in a 5-gal. bucket and then stir in an alkaline-base neutralizer. Using a stiff-bristle broom, make sure the concrete surface is completely covered with the solution. Continue to sweep until the fizzing stops.

Use a garden hose with a pressure nozzle or, ideally, a pressure washer in conjunction with a stiff-bristle broom to thoroughly rinse the concrete surface. Rinse the surface two to three times. Reapply the acid (repeat steps 5, 6, 7, and 8).

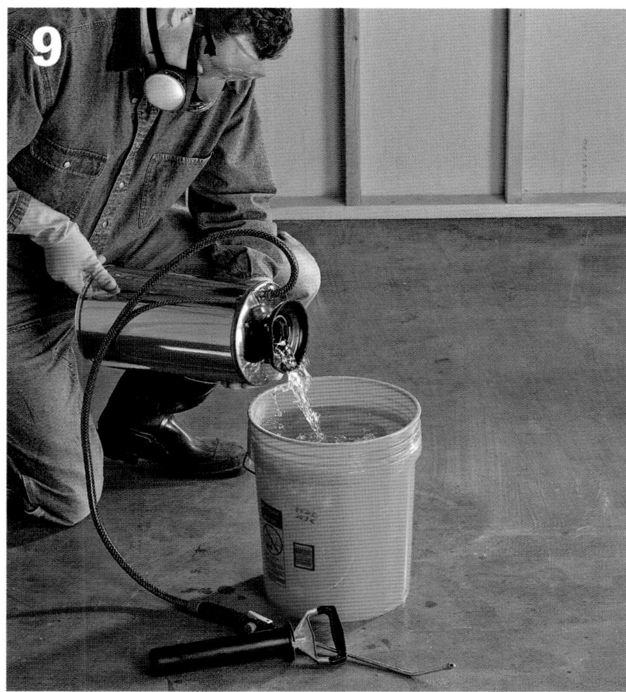

If you have any leftover acid, you can make it safe for your septic system by mixing more alkaline solution in the 5-gal. bucket and carefully pouring the acid from the spray unit into the bucket until all of the fizzing stops.

Use a wet/dry vacuum to clean up the mess. Some sitting acids and cleaning solutions can harm local vegetation, damage your drainage system, and are just plain environmentally unfriendly. Check your local disposal regulations for proper disposal of the neutralized spent acid.

To check for residue, rub a dark cloth over a small area of concrete. If any white residue appears, continue the rinsing process. Check for residue again.

Let the concrete dry for at least 24 hours and sweep up dust, dirt, and particles leftover from the acid-etching process. Your concrete should now have the consistency of 120-grit sandpaper and be able to accept concrete sealants.

How to Stain a Concrete Floor

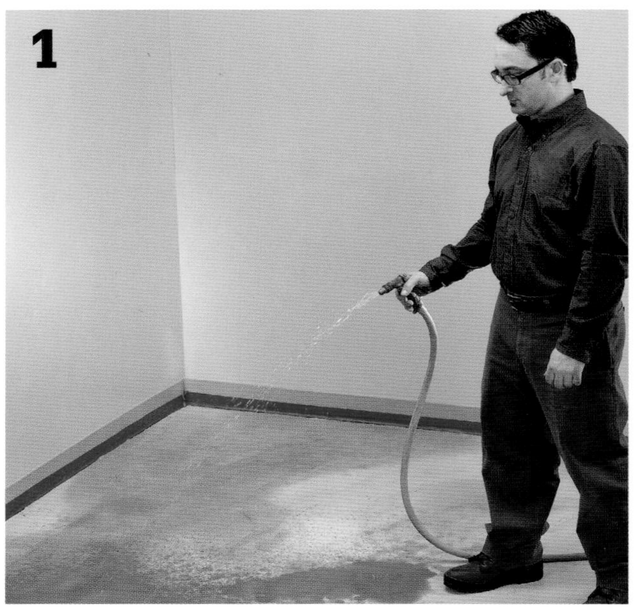

Thoroughly clean the entire floor (see page 165). Use painter's tape and plastic sheeting to protect any areas that won't be stained, as well as surrounding walls and other surfaces. Test the spray of your garden sprayer using water: It should deliver a wide, even mist.

Dampen the floor with water using a garden sprayer. Mop up any pooled water, but make sure the entire floor is damp. Load sprayer with stain, and then apply the stain evenly in a circular motion until the concrete is saturated. Let the floor dry.

Remove the etching residue by soaking the floor with water and scrubbing vigorously with a stiff-bristled brush. As you work, clean up the liquid with a wet/dry vacuum. Dispose of the waste liquid safely, according to local regulations.

When the floor has dried completely (at least 18 to 24 hours), begin applying the sealer along the edges and in any hard-to-reach areas using a paintbrush.

Using a $\frac{3}{8}$" nap roller, apply the sealer in 2 × 6-ft. sections, maintaining a wet edge to prevent lap marks. If the sealer rapidly sinks into the concrete, apply a second coat after 2 hours. Let the floor dry for 18 to 24 hours before allowing light foot traffic and 72 hours before heavy use.

How to Paint a Concrete Floor

If you expect to use more than one container of paint, open them all and mix them together for a uniform color. You do not need to thin a paint for use on a floor, unless you use a sprayer that requires thinned paint.

Using a nylon brush, such as a 2½" sash brush, cut in the sides and corners with primer. This creates a sharp, clean edge. Start this way for the top coat as well.

Using a roller pad with the nap length recommended by the manufacturer, apply a primer coat to the surface. Start at the corner farthest away from the door and back up as you work. Allow the primer to dry for at least 8 hours.

With a clean roller pad, apply the first top coat. Make the top coat even but not too thick, and then let it dry for 24 hours. If you choose to add another top coat, work the roller in another direction to cover any thin spots. Let the final coat dry another day before you walk on it.

Stucco Finish

Prized for its weather resistance, durability, and timeless beauty, stucco has long been one of the most popular exterior wall finishes. As a building material, stucco is essentially an exterior plaster made of Portland cement, sand, and water. Other ingredients may include lime, masonry cement, and various special additives for enhancing properties like crack resistance, workability, and strength. With a few exceptions, stucco is applied much as it has been for centuries—a wet mix is troweled onto the wall in successive layers with the final coat providing the finished color and any decorative surface texture desired.

The two traditional stucco systems are the three-coat system used for standard wood-framed walls and the two-coat system used for masonry walls such as brick, poured concrete, and concrete block. And today, there's a third process—the one-coat system—which allows you to finish standard framed walls with a single layer of stucco, saving you money and considerable time and labor over traditional three-coat applications. Each of these systems is described in detail on the opposite page.

The following pages show you an overview of the materials and basic techniques for finishing a wall with stucco. While cladding an entire house or addition is a job for professional masons, smaller projects and repair work can be much more doable for the less experienced. Fortunately, all the stucco materials you need are available in a dry preblended form, so you can be sure of getting the right blend of ingredients for each application. During your planning, consult with the local building department to learn about requirements for surface preparation, fire ratings for walls, control joints, drainage, and other critical factors.

Stucco is one of the most durable and low-maintenance wall finishes available, but it requires getting each stage of the installation right, as well as the mix of the stucco itself.

Tools & Materials ▸

- Aviation snips
- Stapler
- Hammer
- Level
- Cement mixer
- Wheelbarrow
- Mortar hawk
- Square-end trowel
- Raking tool
- Darby or screed board
- Wood float
- Grade D building paper
- Texturing tools
- Heavy-duty staples
- 1½" galvanized roofing nails
- Self-furring galvanized metal lath (min. 2.5 lb.)
- Metal stucco edging
- Flashing
- Stucco mix
- Nonsag polyurethane sealant
- Work gloves

Base Coat Stucco Calculator ▸

Square feet	10	25	100	300	500
⅜" thick – # of 80-lb. bags	1	1	4	12	19
½" thick – # of 80-lb. bags	1	2	6	16	26
¾" thick – # of 80-lb. bags	1	2	8	24	38

All yields are approximate and do not allow for waste or uneven substrate, etc.

Stucco Systems ▸

Three-coat stucco is the traditional application for stud-framed walls covered with plywood, oriented strand board (OSB), or rigid foam insulation sheathing. It starts with two layers of Grade D building paper for a moisture barrier. The wall is then covered with self-furring, expanded metal lath fastened to the framing with galvanized nails.

The scratch coat is the first layer of stucco. It is pressed into the lath, and then smoothed to a flat layer about ⅜" thick. While still wet, the stucco is "scratched" with a raking tool to create horizontal keys for the next layer to adhere to.

The brown coat is the next layer. It's about ⅜"-thick and brings the wall surface to within ⅛" to ¼" of the finished thickness. Imperfections here can easily telegraph through the thin final coat, so the surface must be smooth and flat. To provide tooth for the final layer, the brown coat is finished with a wood float for a slightly roughened texture.

The finish coat completes the treatment, bringing the surface flush with the stucco trim pieces and providing the color and decorative texture, if desired. There are many options for texturing stucco; a few of the classic ones are shown on page 255.

Two-coat stucco is the standard treatment for masonry walls. This system is the same as a three-coat treatment but without a scratch coat. The base coat on a two-coat system is the same as the brown coat on a three-coat system. For the base coat to bond well, the masonry surface must be clean, unpainted, and sufficiently porous. You can test this by spraying water onto the surface: If the water beads and runs down the wall, apply bonding adhesive before applying the base coat, or you can fasten self-furring metal lath directly to the wall, and then apply a full three-coat stucco treatment.

One-coat stucco is a single-layer system for finishing framed walls prepared with a waterproof barrier and metal lath (as with a three-coat system). This treatment calls for one-coat, fiberglass-reinforced stucco, a special formulation that contains ½" alkali-resistant fiberglass fiber and other additives to combine high-performance characteristics with greatly simplified application. This stucco is applied in a ⅜- to ⅝"-thick layer using standard techniques. QUIKRETE One Coat Fiberglass Reinforced Stucco meets code requirements for a one-hour firewall over wood and form systems.

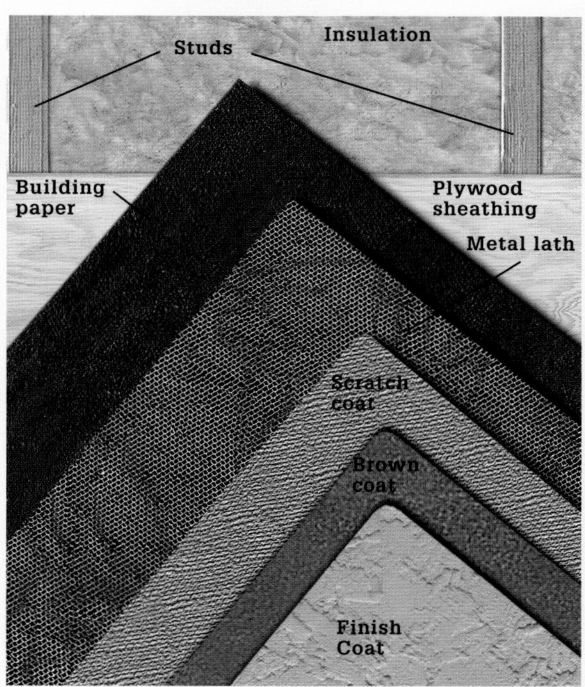

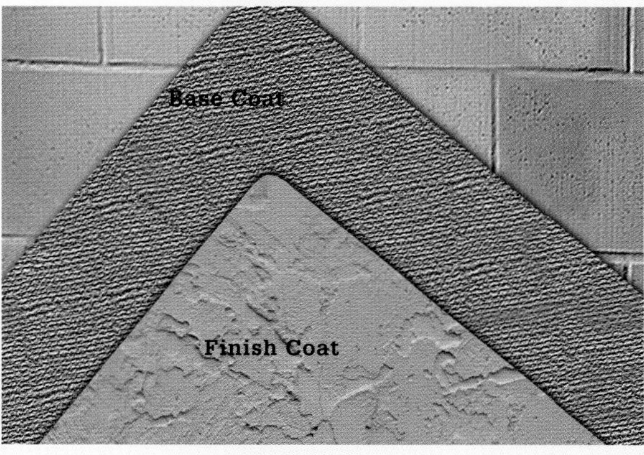

Texturing with Stucco ▸

Finding the right blend of ingredients and mixing to the proper consistency are critical to the success of any stucco project. Premixed stucco eliminates the guesswork by giving you the perfect blend in each bag, along with mixing and curing instructions for a professional-quality job. All of the stucco products shown here are sold in complete form, meaning all you do is add water before application. Be sure to follow the mixing and curing instructions carefully for each product.

Base coat stucco: Use this premixed stucco for both the scratch and brown coats of a three-coat application or for the base coat of a two-coat system. You can apply the mixed stucco with a trowel or an approved sprayer. Available in 80-pound bags in gray color. Each bag yields approximately 0.83 cubic foot or an applied coverage of approximately 27 sq. ft. at ⅜" thickness.

Finish coat stucco: Use this stucco for the finish coat on both three-coat and two-coat systems. You can also use it to create a decorative textured finish over one-coat stucco. Apply finish coat stucco to a minimum thickness of ⅛", and then texture the surface as desired. Available in gray and white for achieving a full range of colors (see below). Coverage of 80-pound bag is approximately 70 square feet at ⅛" thickness.

One-coat, fiberglass-reinforced stucco: Complete your stucco application in one step with this convenient all-in-one stucco mix. You can texture the surface of the single layer or add a top coat of finish coat stucco for special decorative effects. Available in 80-pound bags. An 80-pound bag covers approximately 25 square feet of wall at ⅜" thickness.

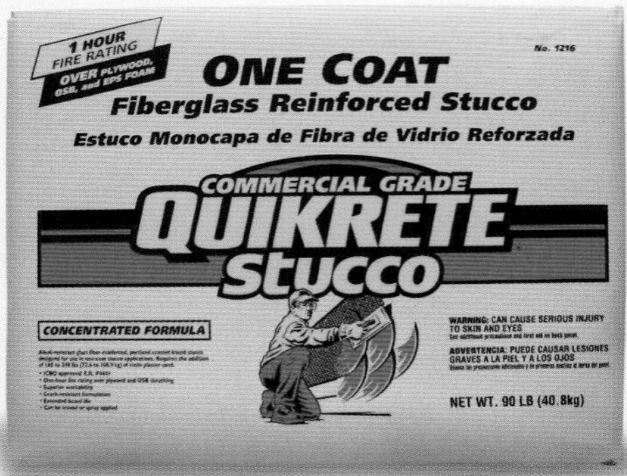

Stucco and mortar color: Available in many colors, stucco and mortar color is a permanent liquid colorant that you blend with the stucco mix before application. Some colors are for use with gray stucco mix, while many others are compatible with white mix. For best results, combine the liquid colorant with the mixing water before adding the dry stucco mix, and then blend thoroughly until the color is uniform.

How to Prepare Framed Walls for Stucco

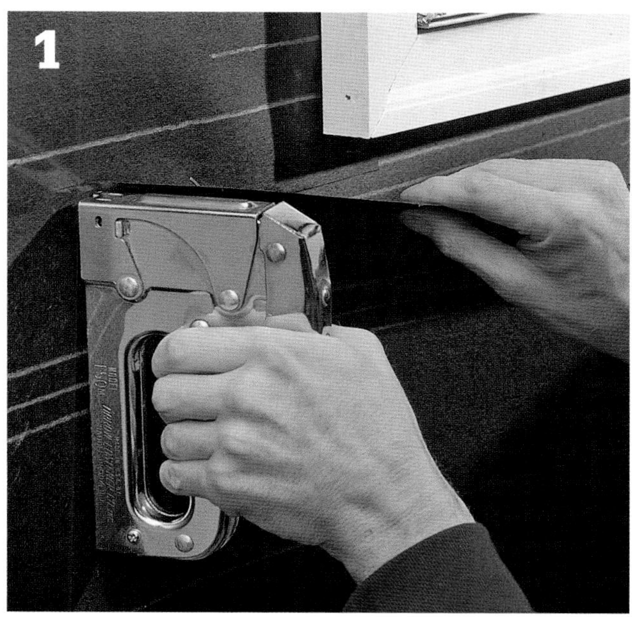

Attach building paper over exterior wall sheathing using heavy-duty staples or roofing nails. Overlap sheets by 4". Usually, two layers of paper are required or recommended; consult your local building department for code requirements in your area.

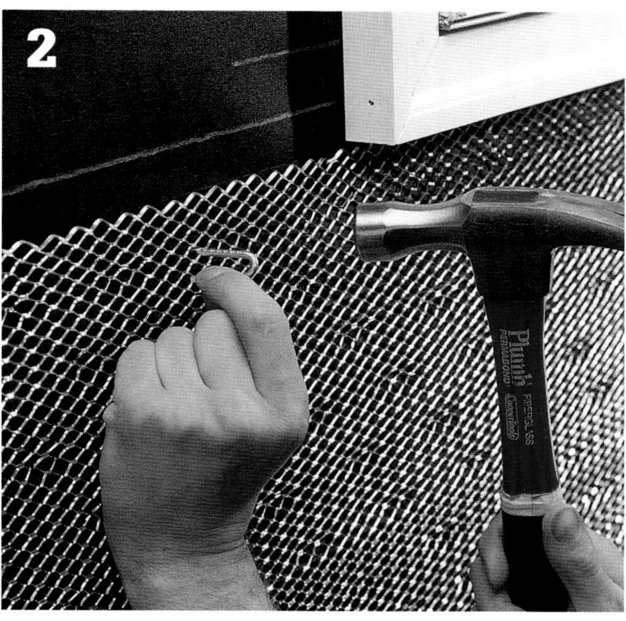

Install self-furring expanded metal lath over the building paper with staples or 1½" galvanized roofing nails (don't use aluminum nails) driven into the wall studs every 6". Overlap sheets of lath by 1" on horizontal seams and 2" on vertical seams. Install the lath with the rougher side facing out.

Install metal edging for clean, finished lines at vertical edges of walls. Install casing bead along the top of stuccoed areas and weep screed (or drip screed) along the bottom edges, as applicable. Make sure all edging is level and plumb, and fasten it with galvanized roofing nails. Add flashing as needed over windows and doors.

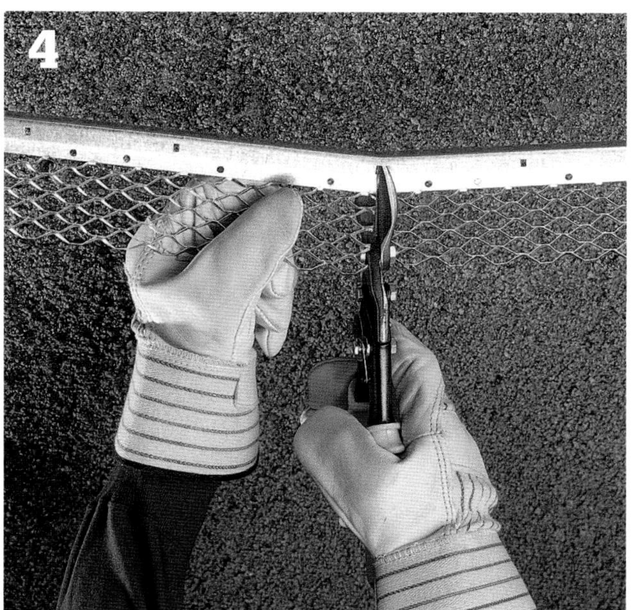

Use aviation snips to trim sheets of lath or cut edging materials to length. Cut lath and edging can be very sharp, so always wear gloves when working with these materials.

How to Finish Walls with Stucco

For a three-coat system, mix the stucco to a trowel-able consistency and apply it with a square trowel, working from the bottom up. Press the stucco into the lath, and then screed and smooth the surface for a uniform thickness. When the coat hardens enough to hold a finger impression, scratch ⅛"-deep horizontal grooves into the surface with a raking tool.

After moist-curing the scratch coat for 24 to 48 hours, mix stucco for the brown coat (or base coat for two-coat system) and apply it in a ⅜"-thick layer. Use a straight board or a darby to screed the surface so it's flat and even. When the stucco has lost its sheen, float it with a wood float to roughen the surface. Moist-cure the coat for 48 hours as directed.

Variation: For a one-coat application, mix a fiber-reinforced one-coat stucco and apply it in a ⅜ to ⅝"-thick layer, working from the bottom up and forcing it in to completely embed the lath. Screed the surface flat with a darby or board. When the surface loses its sheen, finish trowel or texture the surface as desired. Cure the coat as directed. Seal all joints around building elements with polyurethane sealant.

Mix the finish coat and apply it in a ⅛"-thick (minimum) layer, working from the bottom up. Complete large sections or entire walls at one time for color consistency. Texture the surface as desired. Cure the coat as directed. Seal all joints around building elements with polyurethane sealant.

How to Finish Stucco

Test the coloring of finish stucco by adding different proportions of colorant and mix. Let the samples dry to see their true finished color. For the application batches, be sure to use the same proportions when mixing each batch.

Mix the finish batch so it contains slightly more water than the scratch and brown coats. The mix should still stay on the mortar hawk without running.

Finish Option: Cover a float with carpet to make an ideal tool for achieving a float-finish texture. Experiment on a small area.

Finish Option: Achieve a wet-dash finish by flinging, or dashing, stucco onto the surface. Let the stucco cure undisturbed.

Finish Option: For a dash-trowel texture, dash the surface with stucco using a whisk broom (left), and then flatten the stucco by troweling over it.

Stone Veneer

Whether you use natural or manufactured veneer, wet each stone, and then apply mortar to the back before pressing it onto the mortared wall. Wetting and mortaring a stone (called buttering) results in maximum adhesion between the stone and the wall. The challenge is to arrange the stones so that large and small stones and various hues and shapes alternate across the span of the wall.

This project is designed for installing veneer stone over plywood sheathing, which has the strength to support layers of building paper. If your walls are covered with fiberboard or any other type of sheathing, ask the veneer manufacturer for recommendations.

Note: Installing from the top down makes cleanup easier since it reduces the amount of splatter on preceding courses. However, manufacturers advise bottom-up installation for some veneers. Read the manufacturer's guidelines carefully before you begin.

Quick Tip ▸

Find the square footage of veneer stone required for your project by multiplying the length by the height of the area. Subtract the square footage of window and door openings and corner pieces. It's best to increase your estimate by 5 to 10 percent to allow for trimming.

Tools & Materials ▸

- Hammer or staple gun
- Drill
- Wheelbarrow
- Hoe
- Square-end trowel
- Circular saw
- Dust mask
- Stiff-bristle brush
- Wide-mouth nippers or mason's hammer
- Level
- Jointing tool
- Veneer stones
- Veneer stone mortar or Type S mortar mix
- Mortar bag and grout bag
- Spray bottle
- Whisk broom
- Expanded galvanized metal lath (diamond mesh, minimum 2.5#)
- 15# building paper
- Mortar color (optional)
- 1½" (minimum) galvanized roofing nails or heavy-duty staples
- 2 × 4 lumber
- Angle grinder with diamond blade
- Eye protection and work gloves

A splash of manufactured veneer stone, with its variations in color, tone and shape, can set your home apart from many of today's cookie-cutter designs.

How to Finish Walls with Stone Veneer

Cover the wall with building paper, overlapping seams by 4". Nail or staple lath every 6" into the wall studs and midway between studs. Nails or staples should penetrate 1" into the studs. Paper and lath must extend at least 16" around corners where veneer is installed.

Stake a level 2 × 4 against the foundation as a temporary ledger to keep the bottom edge of the veneer 4" above grade. The gap between the bottom course and the ground will reduce staining of the veneer by plants and soil.

Spread out the materials on the ground so you can select pieces of varying size, shape, and color, and create contrast in the overall appearance. Alternate the use of large and small, heavily textured and smooth, and thick and thin pieces.

Mix a batch of veneer stone mortar that's firm but still moist. Mortar that's too dry or too wet is hard to work with and may fail to bond properly.

(continued)

Use a square-end trowel to press a ⅜ to ½" layer of mortar into the lath called the scratch coat. To ensure that mortar doesn't set up too quickly, start with a 5-sq.-ft. area. Before the mortar is set, use a brush or rake to roughen the surface. Allow to set hard before moving onto the next step. *Tip: Mix in small amounts of water to retemper mortar that has begun to thicken.*

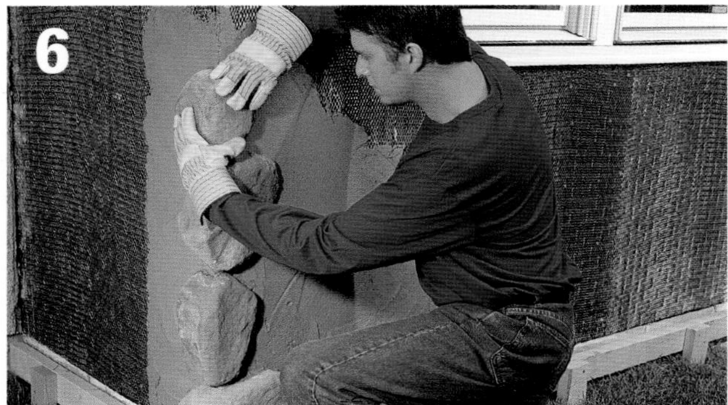

Install corner pieces first, alternating long and short legs. Dampen and apply mortar to the back of each piece, and then press it firmly against the scratch coat so some mortar squeezes out. Joints between stones should be no wider than ½" and should remain as consistent as possible across the wall.

Quick Tip ▸

Polymer-modified veneer stone mortar is recommended for drystack stone applications. Drystack stone is primarily bonded on one edge, requiring twice the bond strength of regular mortar.

Set the first course of block into mortar, following the basic techniques shown on pages 126 to 129. Cut blocks as needed for the door openings. Lay the second course, offsetting the joints with the first course in a running-bond pattern.

If mortar becomes smeared on a stone, remove it with a whisk broom or soft-bristle brush after the mortar has begun to dry. Never use a wire brush or a wet brush of any kind.

Use wide-mouth nippers or a mason's hammer to trim and shape pieces to fit. Do your best to limit trimming so each piece retains its natural look.

You can hide cut edges that are well above or below eye level simply by rotating a stone. If an edge remains visible, use mortar to cover. Let the mortar cure for 24 hours, and then remove the 2 × 4 and stakes, taking care not to dislodge any stones.

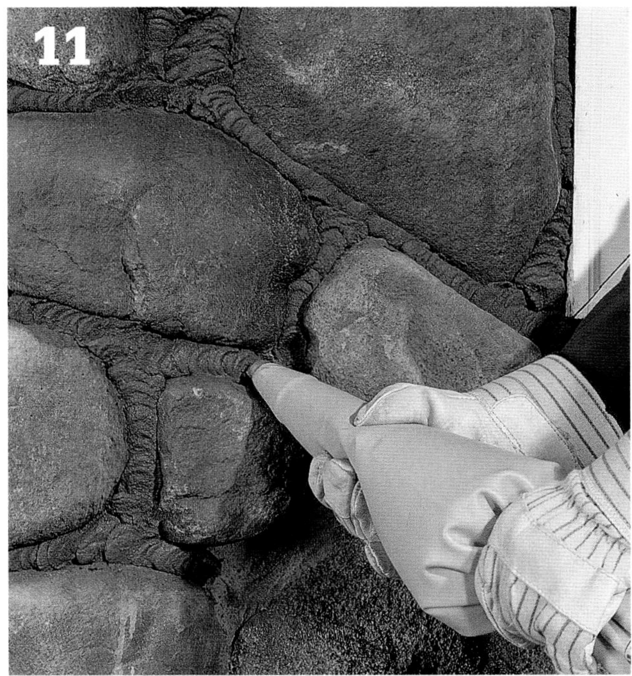

Once the wall is covered in veneer, fill in the joints using a mortar bag and tuck-pointing mortar. Take extra care to avoid smearing the mortar. You can tint the tuck-pointing mortar to complement the veneer.

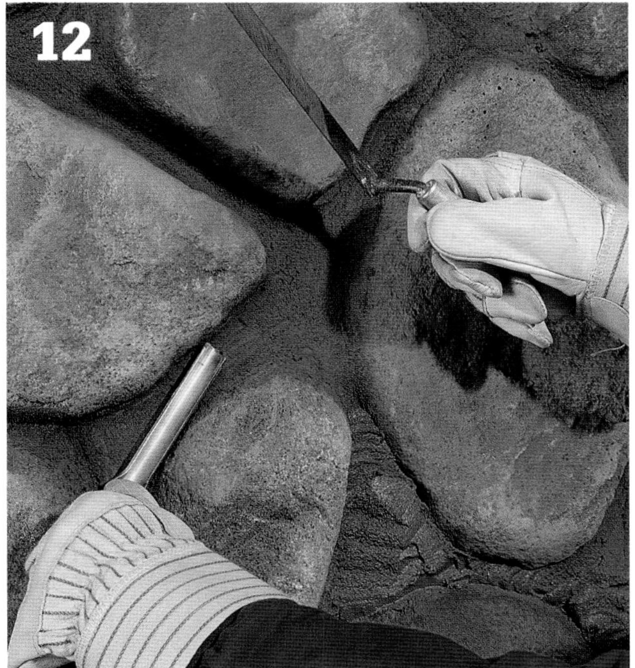

Smooth the joints with a jointing tool once the mortar is firm. Once the mortar is dry to the touch, use a dry whisk broom to remove loose mortar—water or chemicals can leave permanent stains.

Variation: Cast Veneer Stone

Cast veneer stones are thin synthetic masonry units that are applied to building walls to imitate the appearance of natural stone veneer. They come in random shapes, sizes, and colors, but they are scaled to fit together neatly without looking unnaturally uniform. Outside corner stones and a sill block (used for capping half-wall installations) are also shown here.

How to Install Cast Veneer Stone

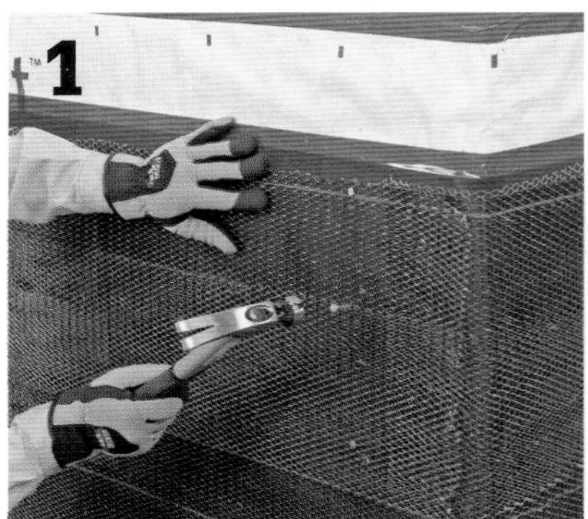

Prepare the wall. Veneer stones can be applied to a full wall or as an accent on the lower portion of a wall. A top height of 36 to 42" looks good. A layer of expanded metal lath (stucco lath) is attached over a substrate of building paper.

Apply a scratch coat. The wall in the installation area should be covered with a ½- to ¾"-thick layer of mortar. Mix one part Type N mortar to two parts masonry sand and enough water to make the consistency workable. Apply with a trowel, and let the mortar dry for 30 minutes. Brush the surface with a stiff-bristle brush.

Test layouts. Uncrate large groups of stones and dry-lay them on the ground to find units that blend well together in shape as well as in color. This will save an enormous amount of time as you install the stones.

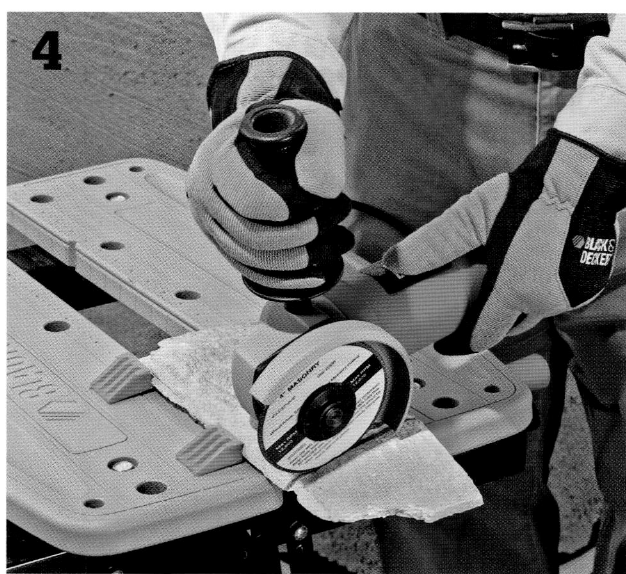

Cut veneer stones, if necessary, by scoring with an angle grinder and diamond blade along a cutting line. Rap the waste side of the cut near the scored line with a mason's hammer or a maul. The stone should fracture along the line.
Try to keep the cut edge out of view as much as you can.

Apply the stones. Mix mortar in the same ratios as in step 2, but instead of applying it to the wall, apply it to the backs of the stones with a trowel. A ½"-thick layer is about right. Press the mortared stones against the wall in their position. Hold them for a few second so they adhere.

Fill the gaps between stones with mortar once all of the stones are installed and the mortar has had time to dry. Fill a grout bag (sold at concrete supply stores) with mortar mixture and squeeze it into the gaps. Once the mortar sets up, strike it smooth with a jointing tool.

Option: Install sill blocks. These are heavier and wider than the veneer block so they require some reinforcement. Attach three 2 × 2" zinc-coated L-brackets to the wall for each piece of sill block. Butter the backs of the sill blocks with mortar and press them in place, resting on the L-brackets. Install metal flashing first for extra protection against water penetration.

Mortarless Brick Veneer Siding

An interesting new siding product is now available that mimics the appearance and durability of classic brick, but that installs as easily as any other siding material. Mortarless brick veneer systems use stackable bricks to create an appealing façade on wood, steel, or concrete structures. The high-strength concrete bricks are long-lasting—manufacturers offer warranties up to 50 years. And because brick veneer does not require mortar, installation is well within the capabilities of interested homeowners.

Veneer bricks are available in three and four inch heights, and they are either eight or nine inches long, depending on the producer. Bricks weigh approximately five pounds each and add three and a quarter inches to the face of walls. While veneer brick systems can be used in both new construction and remodel projects, application is restricted due to the added load: up to 30 feet high on standard wood framed walls. Consult with a professional builder or structural engineer for walls taller than 30 feet, as well as sections of wall above roofs.

When planning a brick veneer siding project, it's best to work with the manufacturer or a local dealer to determine material availability in your area. To accurately estimate materials, you'll need to know the total surface area to be covered with veneer brick, the width of each wall and opening for starter strip, and the length of each corner for quantity of corner strip and corner block.

Prior to installation, make sure the framing and wall substrate is sound and the house is adequately insulated. Extend all plumbing and electrical pipes, boxes, and meters to accommodate the additional thickness created by the veneer brick and furring strips.

Brick veneer siding attaches to your house with mechanical fasteners, so you can achieve the appeal of brick without the mess of mortar.

The following pages discuss the installation of brick veneer siding on standard wood framed walls. All openings require extra support in the form of three-quarter-inch plywood lintels. Lintel size is determined by the width of the opening and the brick installation method over the opening (soldier coursing shown here). Contact the manufacturer or product producer (see Resources, page 314) for information regarding lintel sizing, as well as installation of veneer brick on other framing styles.

Tools & Materials ▸

- Tape measure
- Chalk line
- 4-ft. level
- Utility knife
- Circular saw
- Miter saw with diamond masonry-cutting blade or wet masonry saw
- Hammer drill with masonry bits
- Rubber mallet
- Cordless drill with various drivers
- Caulk gun
- Eye protection
- Work gloves
- Safety glasses
- Dust mask
- Earplugs
- ¾" plywood
- Furring strips
- Flashing
- Self-adhesive waterproof membrane (1 × 3, 1 × 4, 1 × 6)
- #10 × 2½" and #10 × 4" corrosion-resistant wood screws
- Scrap 2 × 4
- Outside corner strips
- Starter strips
- Veneer bricks
- Outside corner blocks
- Inside corner blocks
- Window sill block
- Construction adhesive
- Exterior-grade caulk

Installing Brick Veneer Siding

Brick veneer is stacked in courses with staggered joints, much like traditional brick. However, the first course of the mortarless system is installed on a starter strip and fastened with corrosion-resistant screws to 1 × 3 furring strips at each stud location. Bricks are then fastened every fourth course thereafter. At outside corners, a specialty strip is fastened to 1 × 4 furring strips. Corner blocks for both outside and inside corners are secured with screws and construction adhesive.

Before installing brick veneer siding, make sure all openings are properly sealed. For best results, use a self-adhesive waterproof membrane. Install the bottom strip first, and then the side strips so they overlap the bottom strip. Place the top strip to overlap the sides. Install drip-edge flashing where appropriate.

Cut veneer brick using a miter saw with a diamond blade or a wet masonry saw. When cutting brick, protect yourself with heavy work gloves, safety glasses, earplugs, and a dust mask. See pages 116 to 117 for more brick cutting tips.

Predrill holes in bricks that require fastening using a hammer drill with a $\frac{3}{16}$" masonry bit. Position the brick face-up on the ground and secure with your foot. Drill through the notch in the top portion of the brick, holding the drill bit at 90° to the ground.

How to Install Mortarless Brick Veneer Siding

Snap a level chalk line ¾" above the foundation on each wall of the house. Align the bottom end of furring strips above the chalk line. Fasten 1 × 3 furring strips at each stud location with #10 × 2½" corrosion-resistant wood screws. Install 1 × 4 furring strips at outside corners and 1 × 6s at inside corners.

For each opening, cut ¾" plywood lintels to a size of 15" high × 12" longer than the width of the opening. Center the lintel above the opening so 6" extends beyond each side of the frame, and fasten to framing with #10 screws. Install an aluminum drip-edge above the window frame, and then wrap the lintel and flashing with a strip of self-adhesive waterproof membrane.

At outside corners, position the first section of corner strip 2" above the chalk line. Plumb the strip using a 4-ft. level, and then fasten to the framing with #10 × 4" screws every 10" on alternate sides.

Position the starter strip at the chalk line with the flange beneath the ends of the furring strips. Do not overlap corner strips. At inside corners, cut back the starter strip to so it falls 3½" short of adjacent walls. Level the strip, and then secure to the framing with #10 × 4" corrosion-resistant wood screw at each furring location.

Predrill a hole at the notch in each corner block. Drill holes at a 30° angle using a hammer drill and a 3/16" masonry bit. Place a veneer brick on the starter strip for reference, and then slide the first corner block down the corner strip and position it so the bottom edge falls ½" below the bottom edge of the brick. Fasten the brick to the strip with #10 × 2½" wood screws.

Continue to install corner blocks using #10 × 2½" screws and construction adhesive between courses. For the top of the corner, measure the remaining length and cut a piece of corner strip to size. Fasten blocks to this loose length, cutting the final piece to size if necessary (see page 189). Secure one last block to the existing corner using construction adhesive, and then fit the new assembly in place and fasten with #10 × 2½" screws.

For inside corners, predrill holes at 30° angles into inside corner blocks. As with outside corners, position the first block so the bottom edge is ½" below the bottom edge of the first course of veneer brick. Fasten the block to the framing with #10 × 4" wood screws. Continue installing blocks with #10 screws and construction adhesive between each course.

To create the best overall appearance, place a row of bricks on the starter strip so they extend past the width of the most prominent opening on each wall. Place a brick on the second course at each end of the opening, so each sit evenly above the joint of two bricks below. Sight down from the edges of the opening's frame and adjust the entire row to find a pattern that yields the least amount of small pieces of brick around the opening.

(continued)

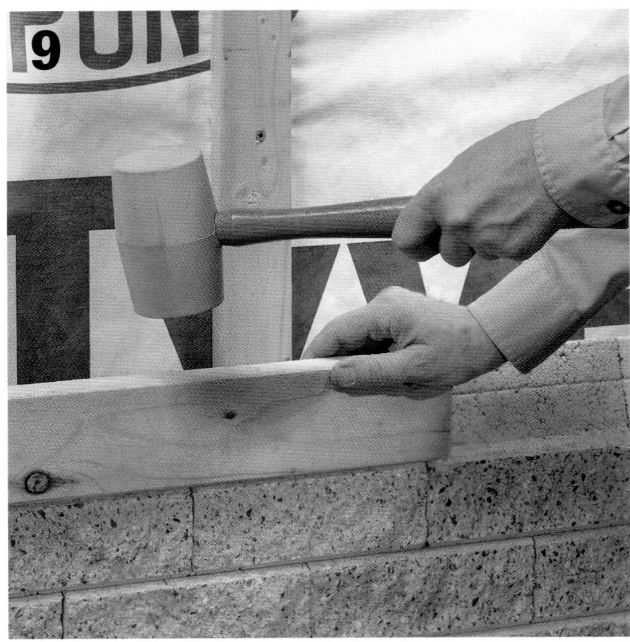

Predrill holes through veneer bricks for the first course (see page 265). Following the established pattern, install bricks on starter strip. At corners, cut bricks to size, so they fit snugly against the blocks. Set bricks using a scrap 2 × 4 and rubber mallet to help maintain consistent course alignment.

At each furring strip, hold bricks flat against the wall and secure to the framing with #10 × 2½" screws. Drive screws until the head touches the brick. Do not over tighten.

Fill the brick courses using bricks from different pallets to blend slight variance in color. Set bricks using a scrap of 2 × 4 and a rubber mallet. Check every fourth course for level before fastening bricks to the framing at each furring strip.

To install sill blocks below the widow, fasten a horizontal 1 × 3 furring strip under the window frame, extending ⅛" longer than the cumulative width of the sill blocks. Install bricks up to the top of the furring strip, cutting to fit as needed, and fasten each with two #10 × 2½" wood screws. Apply construction adhesive along the top of the furring strip and bricks.

Install the sills, angling them downward slightly, and secure with #10 × 4" wood screws, toenailing through the ends or bottom of the sill into the framing. Cut brick filler pieces to bridge the gap between the sill and the last full course of brick; make sure the pieces align with the rest of the course. Install the pieces with construction adhesive. Seal the gap between the window frame and the sill with exterior-grade caulk.

Continue installing brick along the openings, to a height no more than the width of one brick. Cut a piece of starter strip to length, align it with the courses on either side of the opening, and secure to the framing with #10 × 2½" screws. Install a course of bricks on the starter strip, fastening them with #10 screws.

Cut bricks for the soldier course to length, and then install vertically with two #10 × 2½" screws each. For the final brick, cut of the top portion and secure in place with construction adhesive. For a more symmetrical look, place cut bricks in the center of the course.

At the tops of walls, install 1 × 3 horizontal furring strips. Secure the second to the last course of bricks to the framing with #10 × 2½" screws, and then install the last course with construction adhesive. Notch bricks to fit around joists or cut at an angle for gable walls.

Tiling a Concrete Slab

Outdoor tile can be made of several different materials and is available in many colors and styles. A popular current trend is to use natural stone tiles with different shapes and complementary colors, as demonstrated in this project. Tile manufacturers may offer brochures giving you ideas for modular patterns that can be created from their tiles. Make sure the tiles you select are intended for outdoor use.

When laying a modular, geometric pattern with tiles of different sizes, it's crucial that you test the layout before you begin and that you place the first tiles very carefully. The first tiles will dictate the placement of all other tiles in your layout.

You can pour a new masonry slab on which to install your tile patio, but another option is to finish an existing slab by veneering it with tile—the scenario demonstrated here.

Outdoor tile must be installed on a clean, flat, and stable surface. When tiling an existing concrete pad, the surface must be free of flaking, wide cracks, and other major imperfections. A damaged slab can be repaired by applying a one- to two-inch-thick layer of new concrete over the old surface before laying tile.

Note: Wear eye protection when cutting tile and handle cut tiles carefully—the cut edges of some materials may be very sharp.

Tools & Materials ▸

- Tape measure
- Pencil
- Chalk line
- Tile cutter or wet saw
- Tile nippers
- Square-notched trowel
- 2 × 4 padded with carpet
- Hammer
- Grout float
- Grout sponge
- Angle grinder
- Caulk gun
- Tile spacers
- Buckets
- Paintbrush and roller
- Plastic sheeting
- Thinset mortar
- Modular tile
- Grout
- Grout additive
- Grout sealer
- Tile sealer
- Eye protection and work gloves

Stone tiles can be laid as veneer over a concrete patio slab—a very easy way to create an elegant-looking patio.

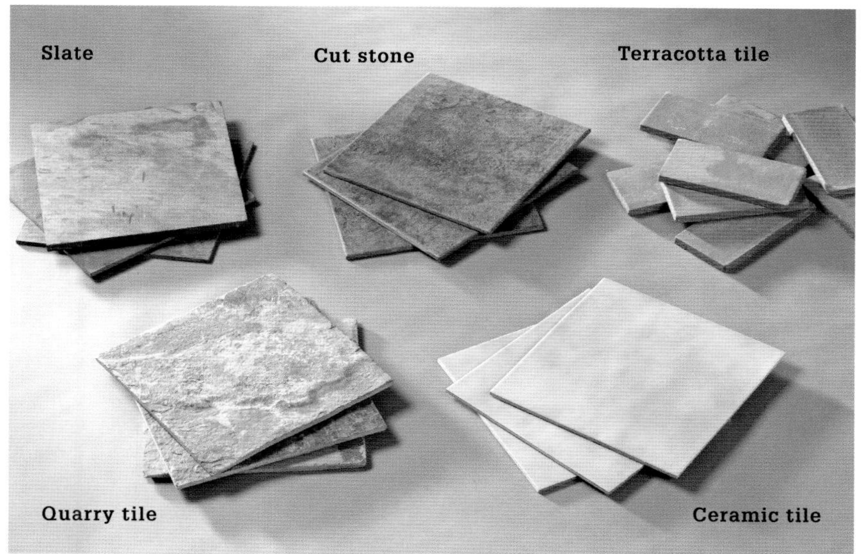

Tile options for landscape installations: Slate and other smooth, natural stone materials are durable and blend well with any landscape but are usually expensive. Quarry tile is less expensive, though only available in limited colors. Exterior-rated porcelain or ceramic tiles are moderately priced and available in a wide range of colors and textures, with many styles imitating the look of natural stone. Terra cotta tile is made from molded clay for use in warmer, drier climates only. Many of these materials require application of a sealer to increase durability and prevent staining and moisture penetration.

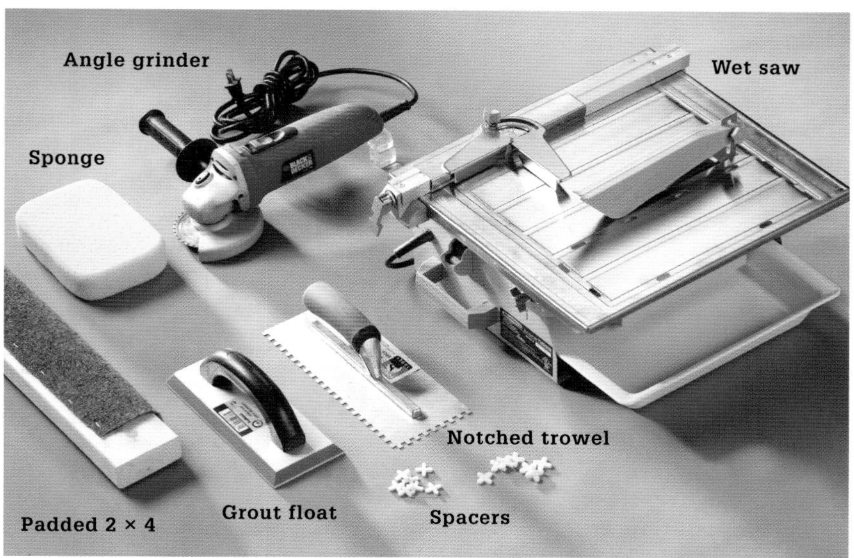

Tools for installing exterior tile include: a wet saw for cutting tile quickly and easily (available at rental centers—make certain to rent one that is big enough for the tile size you install), an angle grinder with a diamond-edged cutting blade (also a rental item) for cutting curves or other complex contours, a trowel with square notches (of the size required for your tile size) for spreading the mortar adhesive, spacers for accurate aligning of tiles and setting consistent joint widths, a straight length of 2 × 4 padded along one edge (carpet pad works well) for helping align tile surfaces, a grout float for spreading grout to fill the joints, and a sponge for cleaning excess grout from tile surfaces.

Materials for installing exterior tile include: latex-modified thinset mortar adhesive that is mixed with water (if you can't find thinset that is latex modified, buy unmodified thinset and mix it with a latex additive for mortar, following manufacturer's directions), exterior-rated grout available in a variety of colors to match the tile you use, grout additive to improve durability, grout sealer to help protect grout from moisture and staining, and tile sealer required for some tile materials (follow tile manufacturer's requirements).

Tips for Evaluating Concrete Surfaces

A good surface is free from any major cracks or badly flaking concrete (called spalling). You can apply patio tile directly over a concrete surface that is in good condition if it has control joints (see below).

A fair surface may exhibit minor cracking and spalling but has no major cracks or badly deteriorated spots. Install a new concrete sub base over a surface in fair condition before laying patio tile.

A poor surface contains deep or large cracks, broken, sunken, or heaved concrete, or extensive spalling. If you have this kind of surface, remove the concrete completely and replace it with a new concrete slab before you lay patio tile.

Tips for Cutting Control Joints in a Concrete Patio

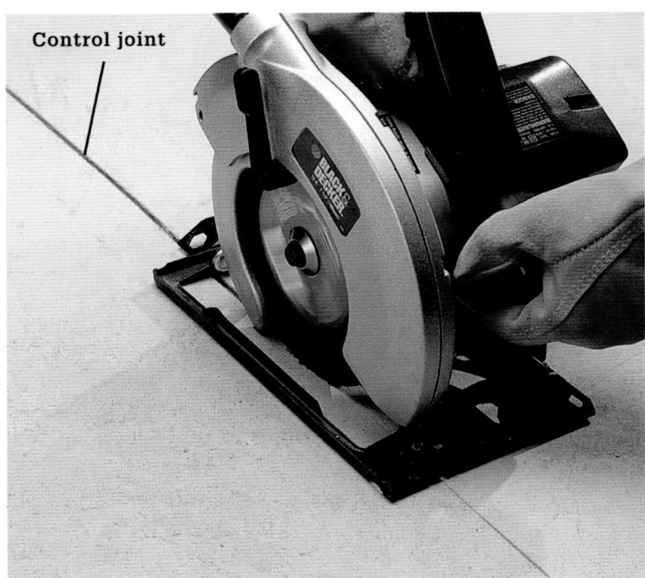

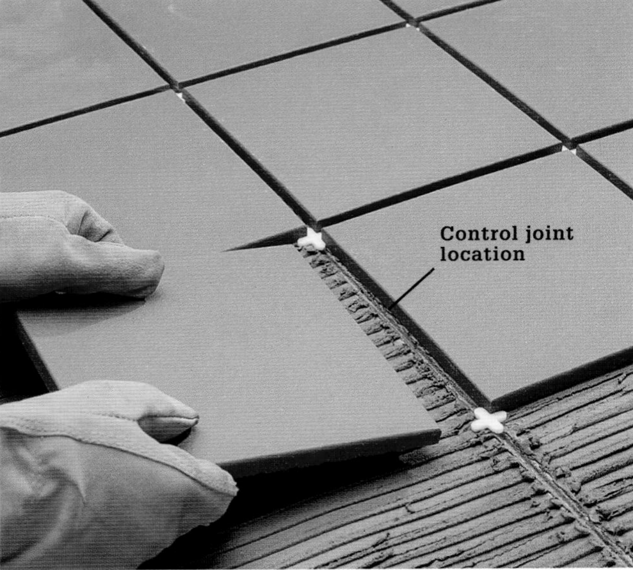

Cut new control joints into existing concrete patios that are in good condition but do not have enough control joints. Control joints allow inevitable cracking to occur in locations that don't weaken the concrete or detract from its appearance. They should be cut every 5 or 6 ft. in a patio. Plan the control joints so they will be below tile joints once the tile layout is established (photo, above right). Use a circular saw with a masonry blade set to ⅜" depth to cut control joints. Cover the saw base with duct tape to prevent it from being scratched.

How to Tile a Patio Slab

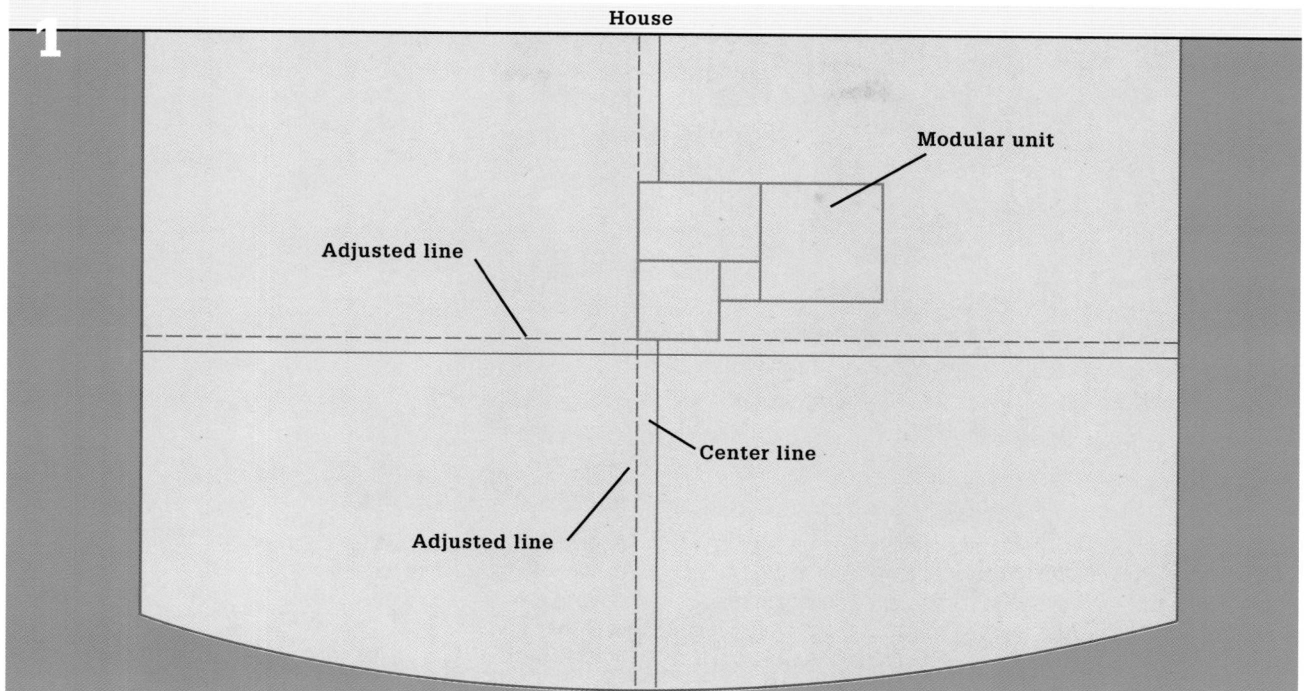

To establish a layout for tile with a modular pattern, you must carefully determine the location of the first tile. On the clean and dry concrete surface, measure and mark a center line down the center of the slab. Test-fit tiles along the line—because of the modular pattern used here, the tiles are staggered. Mark the edge of a tile nearest the center of the pad, and then create a second line perpendicular to the first and test-fit tiles along this line.

Make adjustments as needed so the modular pattern breaks evenly over the patio surface, and it is symmetrical from side to side. You may need to adjust the position of one or both lines. The intersection of the lines is where your tile installation will begin. Outline the position of each group of tiles on the slab.

Variation: To establish a traditional grid pattern, test-fit rows of tiles so they run in each direction, intersecting at the center of the patio. Adjust the layout to minimize tile cutting at the sides and ends, and then mark the final layout and snap chalk lines across the patio to create four quadrants. As you lay tile, work along the chalk lines and in one quadrant at a time.

Following manufacturer's instructions, mix enough thinset mortar to work for about 2 hours (start with 4 to 5" deep in a 5-gallon bucket). At the intersection of the two layout lines, use a notched-edge trowel to spread thinset mortar over an area large enough to accommodate the layout of the first modular group of tiles. Hold the trowel at a 45° angle to rake the mortar to a consistent depth.

Set the first tile, twisting it slightly as you push it into the mortar. Align it with both adjusted layout lines, and then place a padded 2 × 4 over the center of the tile and give it a light rap with a hammer to set the tile.

Position the second tile adjacent to the first with a slight gap between them. Place spacers on end in the joint near each corner and push the second tile against the spacers. Make certain the first tile remains aligned with the layout lines. Set the padded 2 × 4 across both tiles and tap to set. Use a damp cloth to remove any mortar that squeezes out of the joint or gets on tile surfaces. Joints must be at least ⅛"-deep to hold grout.

Lay the remaining tiles of the first modular unit using spacers. Using the trowel, scrape the excess mortar from the concrete pad in areas you will not yet be working to prevent it from hardening and interfering with tile installation.

With the first modular unit set, continue laying tile following the pattern established. You can use the chalk lines for general reference, but they will not be necessary as layout lines. To prevent squeeze-out between tiles, scrape a heavy accumulation of mortar ½" away from the edge of a set tile before setting the adjacent tile.

Tips for Cutting Contours in Tile ▸

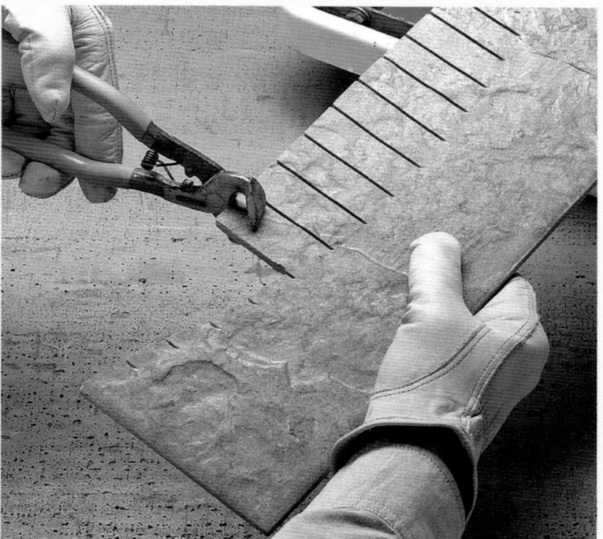

To make convex (above left) or concave (above right) curves, mark the profile of the curve on the tile, and then use a wet saw to make parallel straight cuts, each time cutting as close to the marked line as possible. Use a tile nippers to break off small portions of tabs, gradually working down to the curve profile. Finally, use an angle grinder to smooth off the sharp edges of the tabs. Make sure to wear a particle mask when using the tile saw and wear sturdy gloves when using the nippers.

After installing the tile, remove all the spacers, cover the tiled area with plastic, and let the thinset mortar cure according to the manufacturer's instructions. When tile has fully set, remove the plastic and mix grout using a grout additive instead of water. Grout additive is especially important in outdoor applications because it creates joints that are more resilient in changing temperatures.

Use a grout float to spread grout over an area that is roughly 10 sq. ft. Push down with the face of the float to force grout into the joints, and then hold the float edge at a 45° angle to the tile surfaces and scrape off the excess grout.

Once you've grouted this area, wipe off the grout residue using a damp sponge. Wipe with a light, circular motion—you want to clean tile surfaces but not pull grout out of the joints. Don't try to get the tile perfectly clean the first time. Wipe the area several times, rinsing out the sponge frequently.

11

Once the grout has begun to set (usually about 1 hour, depending on temperature and humidity), clean the tile surfaces again. You want to thoroughly clean grout residue from tile surfaces because it is difficult to remove once it has hardened. Buff off a light film left after final cleaning with a cloth.

Grouting Porous Tiles ▸

Some tiles, such as slate, have highly porous surfaces that can be badly stained by grout. For these tiles, apply grout by filling an empty caulk tube (available at tile stores and some building centers) with grout and apply the grout to the joints with a caulk gun. Cut the tip to make an opening just large enough to allow grout to be forced out. Run the tip down the joint between tiles as you squeeze out the grout. Remove the grout that gets on the tile surface with a wet sponge. You may need to use your finger to force grout into the joint—protect your skin by wearing a rubber glove.

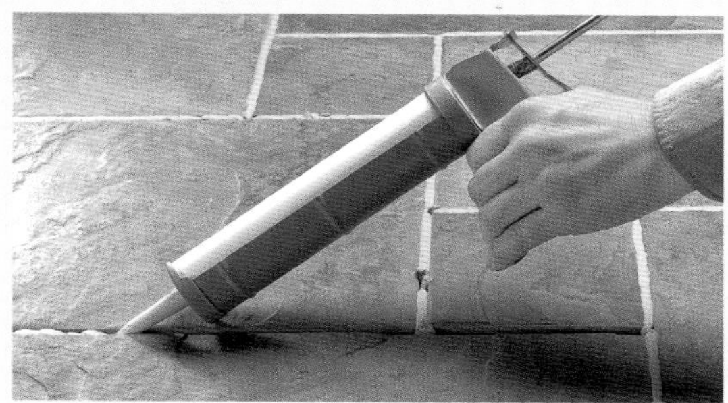

Cover the patio slab with plastic and let the grout cure according to manufacturer's instructions. Once the grout has cured, use a foam brush to apply grout sealer to only the grout, wiping any spillover off of tile surfaces.

Apply tile sealer to the entire surface, using a paint roller. Cover the patio with plastic and allow the sealer to dry completely before exposing the patio to weather or traffic.

Tiling Concrete Steps

In addition to the traditional tricks for improving your home's curb appeal—landscaping, fresh paint, pretty windows—a tiled entry makes a wonderful, positive impression. To be suitable for tiling, stair treads must be deep enough to walk on safely. Check local building codes for specifics, but most require that treads be at least eleven inches deep (from front to back) after the tile is added.

Before you start laying any tiles, the concrete must be free of curing agents, clean, and in good shape. Make necessary repairs and give them time to cure. An isolation membrane can be applied before the tile. This membrane can be a fiberglass sheet or it can be brushed on as a liquid to dry. In either case, it separates the tile from the concrete, which allows the two to move independently and protects the tile from potential settling or shifting of the concrete.

Choose exterior-rated, unglazed floor tile with a skid-resistant surface. Tile for the walking surfaces should be at least one-half-inch thick. Use bullnose tiles at the front edges of treads (as you would on a countertop) and use cove tiles as the bottom course on risers.

Tools & Materials ▸

- Pressure washer
- Masonry trowel
- 4-ft. level
- Carpenter's square
- Straightedge
- Tape measure
- Chalk line
- Tile cutter or wet saw
- Tile nippers
- Square-notched trowel
- Needle-nose plier
- Rubber mallet
- Grout float
- Grout sponge
- Caulk gun
- Latex or epoxy patching compound
- Isolation membrane
- Wet/dry vacuum
- Tile spacers
- Buckets
- Paintbrush and roller
- Plastic sheeting
- Paper towels
- Dry-set (thinset) mortar
- Bonding adhesive
- Field tile
- Bullnose tile
- Grout
- Grout additive
- Latex tile caulk
- Grout sealer
- Tile sealer
- 2 × 4
- Carpet scrap
- Cold chisel or flathead screwdriver
- Wire brush
- Broom
- Eye protection and work gloves

Tiled front steps dramatically upgrade tired-looking, plain concrete. If your concrete is in good condition but you're ready to add curb appeal, this is a perfect DIY solution.

How to Tile Concrete Steps

Use a pressure washer to clean the surface of the concrete. Use a washer with at least 4,000 psi and follow manufacturer's instructions carefully to avoid damaging the concrete with the pressurized spray.

Dig out rubble in large cracks and chips using a small cold chisel or flathead screwdriver. Use a wire brush to loosen dirt and debris in small cracks. Sweep the area or use a wet/dry vacuum to remove all debris.

Fill small cracks and chips with masonry patching compound, using a masonry trowel. Allow the patching compound to cure according to manufacturer's directions.

If damage is located at a front edge, clean it as described above. Place a board in front and block the board in place with bricks or concrete blocks. Wet the damaged area and fill it with patching compound. Use a masonry trowel to smooth the patch and then allow it to cure thoroughly.

Test the surface of the steps and stoop for low spots, using a 4-ft. level or other straightedge. Fill any low spots with patching compound and allow the compound to cure thoroughly.

(continued)

Spread a layer of isolation membrane over the concrete using a notched trowel. Smooth the surface of the membrane, using the flat edge of a trowel. Allow the membrane to cure according to manufacturer's directions.

The sequence is important when tiling a stairway with landing. The primary objective is to install the tile in such a way that the fewest possible cut edges are visible from the main viewing position. If you are tiling the sides of concrete steps, start laying tile there first. Begin by extending horizontal lines from the tops of the stair treads back to the house on the sides of the steps. Use a 4-ft. level.

Mix a batch of thinset mortar with latex bonding adhesive and trowel it onto the sides of the steps, trying to retain visibility of the layout lines. Because the top steps are likely more visible than the bottom steps, start on top and work your way down.

Begin setting tiles into the thinset mortar on the sides of the steps. Start at the top and work your way downward. Try to lay out tile so the vertical gaps between tiles align. Use spacers if necessary.

Wrap a 2 × 4 in old carpet and drag it back and forth across the tile surfaces to set them evenly. Don't get too aggressive here—you don't want to dislodge all of the thinset mortar.

Measure the width of a riser, including the thickness of the tiles you've laid on the step sides. Calculate the centerpoint and mark it clearly with chalk or a high visibility marker.

Next, install the tiles on the stair risers. Because the location of the tops of the riser tiles affects the positioning of the tread and landing tiles, you'll get the most accurate layout if the riser tiles are laid first. Start by stacking tiles vertically against the riser. In some cases, you'll only need one tile to reach from tread to tread. Add spacers. Trace the location of the tread across the back of the top tile to mark it for cutting.

Cut enough tiles to size to lay tiles for all the stair risers. Be sure to allow enough space for grout joints if you are stacking tiles.

Trowel thinset mortar mixed with bonding adhesive onto the faces of the risers. In most cases, you should be able to tile each riser all at once.

Lay tiles on the risers. The bottom tile edges can rest on the tread, and the tops of the top tiles should be flush with or slightly lower than the plane of the tread above.

(continued)

Dry-lay tile in both directions on the stair landing. You'll want to maintain the same grout lines that are established by the riser tiles, but you'll want to evaluate the front-to-back layout to make sure you don't end up with a row of tiles that is less than 2" or so in thickness.

Cut tiles as indicated by your dry run and begin installing them by troweling thinset adhesive for the bullnose tiles at the front edge of the landing. The tiles should overlap the top edges of the riser tiles, but not extend past their faces.

Set the first row of field tiles, maintaining an even gap between the field tiles and the bullnose tiles.

Add the last row of tiles next to the house and threshold, cutting them as needed so they are between ¼ and ½" away from the house.

Install tiles on the stair treads, starting at the top tread and working your way downward. Set a bullnose tile on each side of the center line and work your way toward the sides, making sure to conceal the step side tiles with the tread tiles.

Fill in the field tiles on the stair treads, being sure to leave a gap between the back tiles and the riser tiles that's the same thickness as the other tile gaps.

Let the thinset mortar cure for a few days, and then apply grout in the gaps between tiles using a grout float. Wipe away the grout after it clouds over. In the event of rain, cover with plastic.

After a few weeks, seal the grout lines with an exterior-rated grout sealer.

Select (or have prepared) a pretinted caulk that's the same color as your grout. Fill the gap between the back row of tiles and the house with caulk. Smooth with a wet finger if needed.

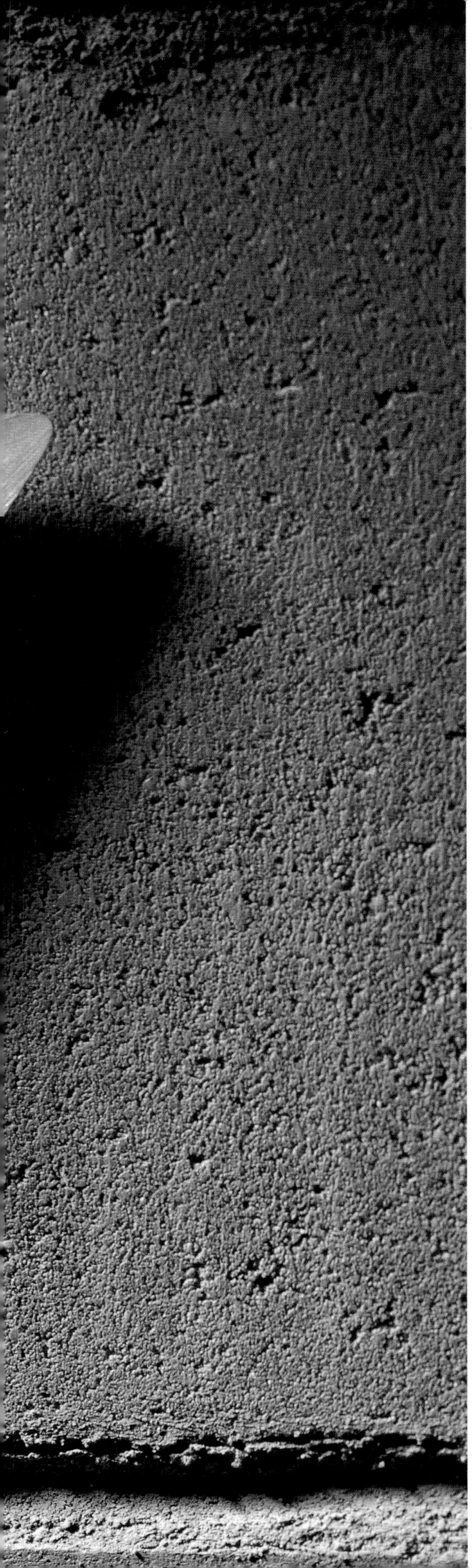

Repair & Maintenance

Concrete, brick, and stucco are among the world's most durable and low-maintenance building materials, but they may need occasional upkeep and repairs to look and perform their best. Because masonry is very rigid (and sometimes because neighboring elements like wood and soil are not), cracking is one of the most common forms of damage. However, unless there's a problem with the underlying soil or gravel base, most cracks have little or no effect on the slab's performance, and repairs are often long-lasting.

In this chapter:

Repairing Concrete

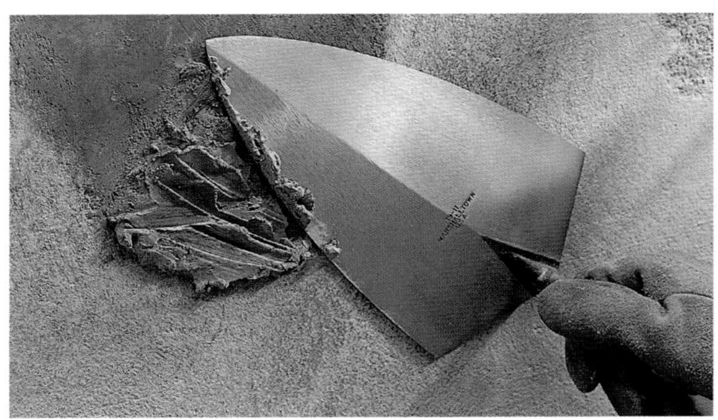

Large and small holes are treated differently when repairing concrete. The best product for filling in smaller holes (less than one half inch deep) is vinyl-reinforced concrete patcher. Reinforced repair products should be applied only in layers that are one quarter inch thick or less. For deeper holes, use sand-mix concrete with an acrylic fortifier, which can be applied in layers up to two inches thick.

Patches in concrete will be more effective if you create clean, backward-angled cuts around the damaged area, to create a stronger bond. For extensive cutting of damaged concrete, it's best to score the concrete first with a circular saw equipped with a masonry blade. Use a chisel and maul to complete the job.

Tools & Materials ▸

- Trowels
- Drill with masonry-grinding disc
- Circular saw with masonry-cutting blade
- Cold chisel
- Hand maul
- Hose
- Paintbrush
- Screed board
- Float
- Scrap lumber
- Vegetable oil or commercial release agent
- Hydraulic cement
- Bonding agent
- Vinyl-reinforced patching compound
- Sand-mix concrete
- Concrete fortifier
- Plastic sheeting
- Eye protection and work gloves

Tool Tip ▸

Use fast-set repair mortar or quick-setting cement with acrylic fortifier for repairing holes and chip-outs in vertical surfaces. Because they set up in just a few minutes, these products can be shaped to fill holes without the need for forms.

How to Patch a Small Hole

Cut out around the damaged area with a masonry-grinding disc mounted on a portable drill (or use a hammer and stone chisel). The cuts should bevel about 15° away from the center of the damaged area. Chisel out any loose concrete within the repair area. Always wear gloves and eye protection.

Dampen the repair area with clean water and then fill it with vinyl concrete patcher. Pack the material in with a trowel, allowing it to crown slightly above the surrounding surface. Then, feather the edges so the repair is smooth and flat. Protect the repair from foot traffic for at least one day and three days from vehicle traffic.

How to Patch a Large Hole

Use a hammer and chisel or a heavy floor scraper to remove all material that is loose or shows any deterioration. Thoroughly clean the area with a hose and nozzle or a pressure washer.

OPTION: Make beveled cuts around the perimeter of the repair area with a circular saw and masonry-cutting blade. The bevels should slant down and away from the damage to create a "key" for the repair material.

Mix concrete patching compound according to the manufacturer's instructions, and then trowel it neatly into the damage area, which should be dampened before the patching material is placed. Overfill the damage area slightly.

Smooth and feather the repair with a steel trowel so it is even with the surrounding concrete surface. Finish the surface of the repair material to blend with the existing surface. For example, use a whisk broom to recreate a broomed finish. Protect the repair from foot traffic for at least one day and three days from vehicle traffic.

Patching Cracks

The materials and methods you should use for repairing cracks in concrete depend on the location and size of the crack. For small cracks (less than one half inch wide), you can use concrete repair caulk for a quick aesthetic fix. Sanded acrylic repair caulks do a good job of matching the color and texture of concrete and stucco surfaces. Larger cracks require concrete repair materials that are fast-setting and high-strength polymer modified compounds that significantly increase the bonding properties and long-term durability of the concrete repair. Cracks that are a result of continual slab movement often cannot be repaired with a rigid concrete repair material. These cracks should be repaired with flexible polyurethane sealants that will elongate with the movement of the flexible concrete crack.

Tools & Materials ▸

Caulk gun
Cold chisel
Pointing trowel
Maul or hammer
Wire brush
Fast-set repair mortar
Quick-setting cement
Acrylic fortifier
Hydraulic waterstop
Vinyl concrete patcher and sand mix
Polyurethane sealant

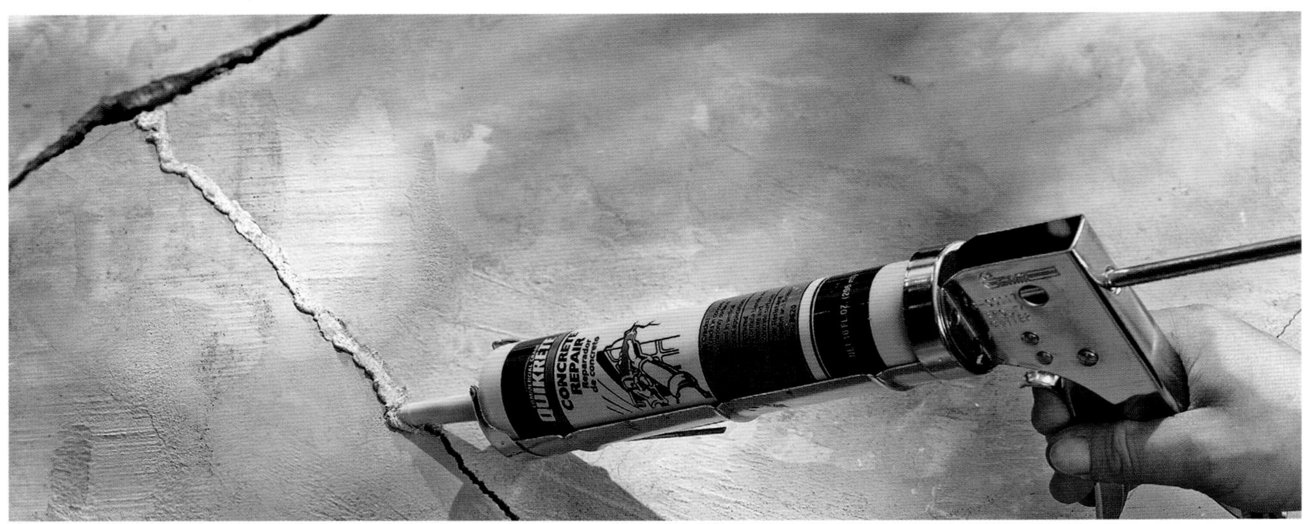

Concrete repair caulk can be forced into small cracks to keep them from expanding. Smooth the caulk after application.

Recommended Crack Preparation

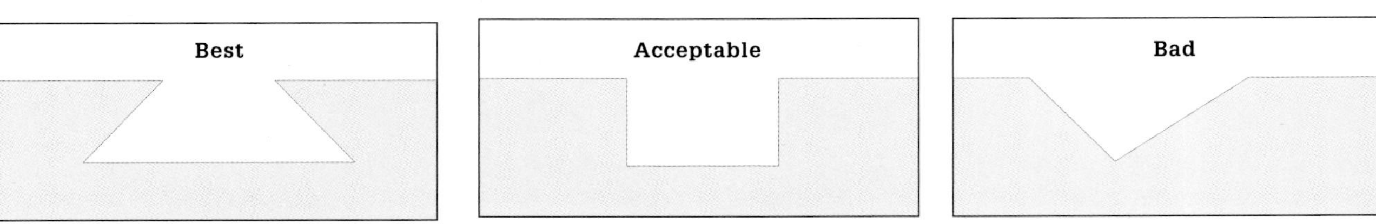

The best way to repair cracks in concrete is to enlarge the crack first by chiseling a keyway along the crack path with a cold chisel. The best holding power for the new patch material is achieved if you chisel in a dovetail shape. A square keyway will also work. A V-shaped keyway will ultimately lead to failure of the repair.

How to Repair Horizontal Cracks

Prepare the crack for the repair materials by knocking away any loose or deteriorating material and beveling the edges down and outward with a cold chisel. Sweep or vacuum the debris and thoroughly dampen the repair area. Do not, however, allow any water to pool.

Mix the repair product to fill the crack according to the manufacturer's instructions. Here, a fast-setting cement repair product with acrylic fortifier is being used. Trowel the product into the crack, overfilling slightly. With the edge of the trowel, trim the excess material and feather it so it is smooth and the texture matches the surrounding surface.

How to Repair Vertical Cracks

Prepare the crack for repair as with a horizontal crack (step 1, above) and then fill the crack with fast-setting repair mortar. The mixture should have a fairly dry consistency so it does not run out of the crack. Overfill the crack slightly and allow the repair material to set up.

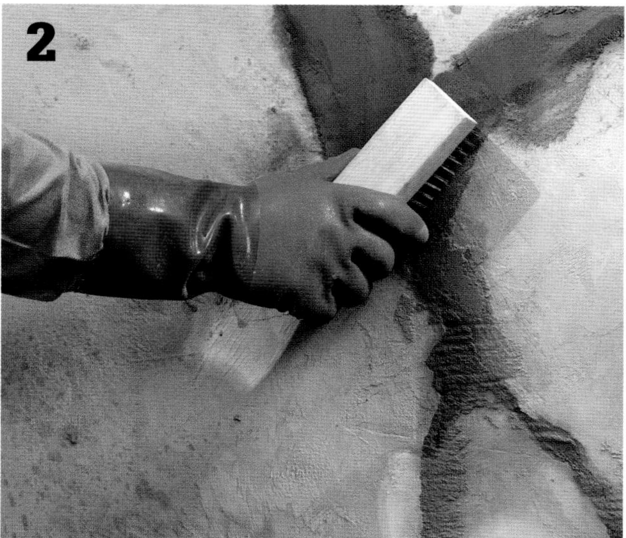

Shape or trim the concrete repair product so it is even with surrounding surface and the textures match. If the material has set too much and is difficult to work, try using a wire attachment on a power drill.

Quick Fixes for Wet Walls

Failing gutters, broken or leaking pipes, condensation, and seepage are the most common causes of basement moisture. If allowed to persist, dampness can cause major damage to concrete basement walls. There are several effective ways to seal and protect the walls. If condensation is the source of the problem, check first that your clothes dryer is properly vented, and install a dehumidifier. If water is seeping in through small cracks or holes in the walls, repair damaged gutters and leaky pipes, and check the grade of the soil around your foundation. Once you've addressed the problem at its source, create a waterproof seal over openings in the basement walls. To stop occasional seepage, coat the walls with masonry sealer. For more frequent seepage, seal the openings and resurface the walls with a water-resistant masonry coating. Heavy-duty coatings, such as surface bonding cement (opposite page), are best for very damp conditions. Thinner brush-on coatings are also available. For chronic seepage, ask a contractor to install a baseboard gutter and drain system.

Remember: To prevent long-term damage, it's necessary to identify the source of the moisture and make repairs both inside and outside your home, so moisture no longer penetrates foundation walls.

Tools & Materials ▸

- Wire brush
- Stiff-bristle paintbrush
- Sponge
- Square-end trowel
- Hydraulic water stop cement
- Heavy-duty masonry coating
- Surface bonding cement

Although most water problems in basements are not caused by cracks in the foundation wall, a large crack should be repaired immediately, especially in a damp basement. To repair it, create a dovetail-shaped keyway with a cold chisel and maul, and then fill the crack with hydraulic repair cement (this product actually cures and hardens when it contacts water).

Tips for Inspecting & Sealing Basement Walls ▸

Paint that is peeling off basement walls usually indicates water seepage from outside that is trapped between the walls and the paint.

Tape a square of aluminum foil to a masonry wall to identify high moisture levels. Check the foil after 24 hours. Beads of water on top of the foil indicate high humidity in the room. Beads of water underneath suggest water seepage through the wall from outside.

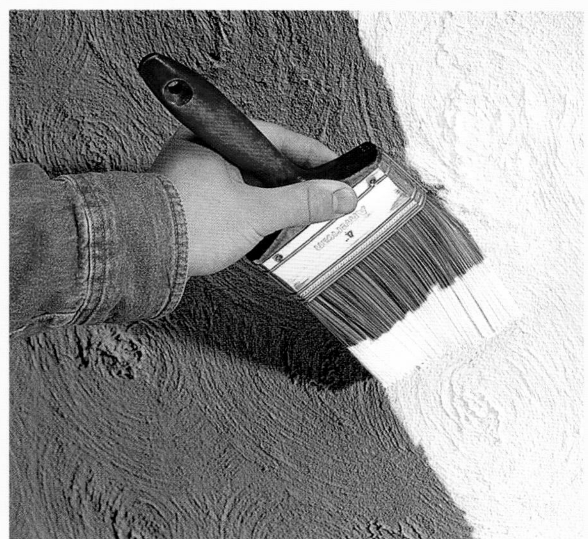

To control minor seepage through porous concrete and masonry, seal walls with a cement-based masonry sealer. Clean the walls and prepare by adding acrylic fortifier to heavy-duty masonry coating. Dampen the walls and apply masonry coating to the walls (including all masonry joints) with a stiff masonry brush in a circular motion.

Resurface heavily cracked masonry walls with a water-resistant masonry coating, such as fiber-reinforced surface bonding cement with acrylic fortifier. Clean and dampen the walls according to the coating manufacturer's instructions, and then fill large cracks and holes with the coating. Finally, plaster a ¼" layer of the coating on the walls using a finishing trowel. Specially formulated heavy-duty masonry coatings are available for very damp conditions.

Renewing an Old Concrete Slab

Over time, exposed concrete surfaces can start to show a lot of wear. Weather, hard use, and problems with the initial pour and finishing are among the most common causes of surface blemishes. But despite a shabby appearance, old concrete is often structurally sound and can last for many more years. So instead of breaking up and replacing an old slab, you can easily renew its surface with concrete resurfacer. With the simple application, your concrete will have a freshly poured look and a protective surface layer that's typically stronger than the slab itself.

Concrete resurfacer is suitable for any size of slab, outdoors or indoors. You can also apply it to vertical surfaces to put a fresh face on steps, curbs, and exposed patio edges. Depending on the condition of the old surface, the new layer can range in thickness from one sixteenth to one quarter inch. For a smooth finish, spread the resurfacer with a squeegee or trowel. For a textured or nonslip surface, you can broom the surface before it dries or use a masonry brush for smaller applications.

Tools & Materials ▸

- 3,500 psi pressure washer
- Steel concrete finishing trowel
- Long-handled squeegee
- Concrete cleaner
- 5-gal. bucket
- ½" drill with paddle mixer
- Duct tape or backer rod
- Stiff-bristle broom
- Concrete resurfacer
- Eye protection and work gloves

Concrete resurfacer offers an easy, inexpensive solution for renewing patios, driveways, paths, steps, and other concrete surfaces that have become chipped and flaked with age.

How to Renew an Old Slab

Thoroughly clean the entire project area. If necessary, remove all oil and greasy or waxy residue using a concrete cleaner and scrub brush. Water beading on the surface indicates residue that could prevent proper adhesion with the resurfacer; clean these areas again as needed.

Wash the concrete with a pressure washer. Set the washer at 3,500 psi, and hold the fan-spray tip about 3" from the surface or as recommended by the washer manufacturer. Remove standing water.

Fill sizeable pits and spalled areas using a small batch of concrete resurfacer—mix about 5 pints of water per 40-lb. bag of resurfacer for a trowelable consistency. Repair cracks or broken slab edges as shown on pages 286 to 287. Smooth the repairs level with the surrounding surface, and let them harden.

On a large project, section off the slab into areas no larger than 100 sq. ft. It's easiest to delineate sections along existing control joints. On all projects, cover or seal off all control joints with duct tape, foam backer rod, or weatherstripping to prevent resurfacer from spilling into the joints.

(continued)

Mix the desired quantity of concrete resurfacer with water, following the mixing instructions. Work the mix with a ½" drill and a mixing paddle for 5 minutes to achieve a smooth, pourable consistency. If necessary, add water sparingly until the mix will pour easily and spread well with a squeegee.

Saturate the work area with water, and then use a squeegee to remove any standing water. Pour the mix of concrete resurfacer onto the center of the repair area or first repair section.

Tool Tip ▸

For improved color consistency, apply a thin, slurry coat of concrete resurfacer to seal the concrete substrate. An additional coat can be applied after two hours.

Spread the resurfacer with the squeegee using a scrubbing motion to make sure all depressions are filled. Then, spread it into a smooth, consistent layer. If desired, broom the surface for a nonslip finish (opposite page). You can also tool the slab edges with a concrete edger within 20 minutes of application. Let the resurfacer cure. Resurface outdoor projects when the temperature will stay above 50°F for 8 hours and the surface won't freeze for at least 24 hours. Cover the area only if necessary to protect it from rain in the first 6 hours of curing (this may affect surface appearance and uniformity). During extreme wind or sun conditions, moist-cure the surface with a water fog-spray twice daily for 24 to 48 hours after application. Let resurfacer cure for 6 hours before allowing foot traffic and 24 hours before vehicle traffic (wait longer during cold weather).

Options for Resurfacer Finishes ▸

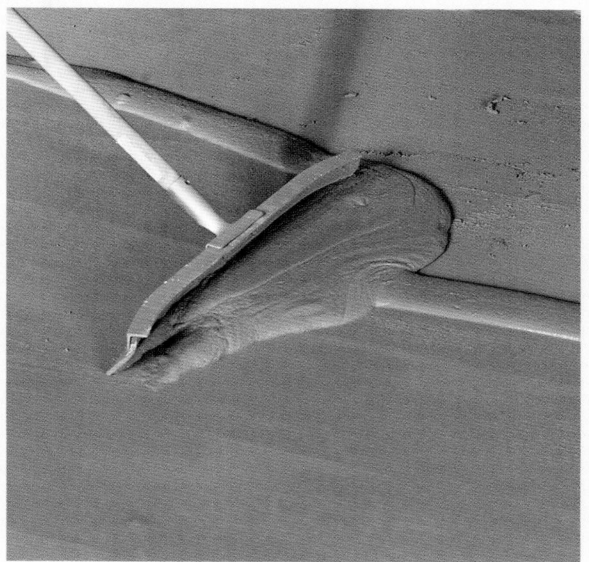

For thicker resurfacing, simply add more layers of resurfacer as needed. Wait until the surface can support foot traffic—typically 2 to 6 hours—before applying the next coat.

Nonslip broomed finish: Within 5 minutes of applying the resurfacer, drag a clean fine-bristled push broom across the surface. Pull the broom backward in a straight line, moving across the entire area without stopping. Repeat in parallel rows until the entire surface is textured.

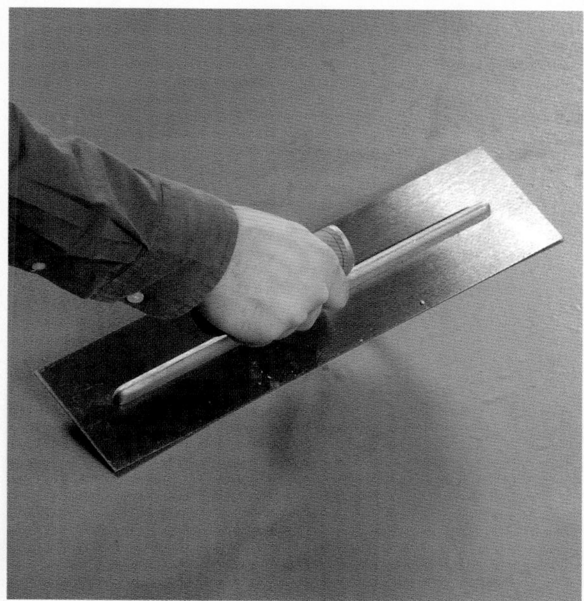

Trowel application: A trowel is handy for resurfacing small areas. Use a stiffer mix for troweling—approximately 5 pints of water per 40-lb. bag of dry mix. Spread and smooth the resurfacer with a steel concrete finishing trowel.

Brush application: Resurface curbs, step risers, and slab edges using a masonry brush. Mix a workably stiff batch of resurfacer, and apply it evenly over the repair area. Finishing the surface with short brush strokes produce a mottled appearance; straight, continuous strokes create a broomed look.

Repairing Steps

Steps require more maintenance and repair than other concrete structures around the house because heavy use makes them more susceptible to damage. Horizontal surfaces on steps can be treated using the same products and techniques used on other masonry surfaces. For vertical surfaces, use quick-setting cement, and shape it to fit.

Damaged concrete steps are an unsightly and unsafe way to welcome visitors to your home. Repairing cracks as they develop not only keeps the steps in a safer and better looking condition, but also prolongs their life.

Tools & Materials ▸

- Trowel
- Wire brush
- Paintbrush
- Circular saw with masonry-cutting blade
- Chisel
- Float
- Edger
- Scrap lumber
- Eye protection
- Vegetable oil or commercial release agent
- Latex bonding agent
- Vinyl-reinforced patching compound
- Quick-setting cement
- Plastic sheeting
- Work gloves

How to Patch a Corner

Clean chipped concrete with a wire brush. Brush the patch area with latex bonding agent.

Mix patching compound with latex bonding agent, as directed by the manufacturer. Apply the mixture to the patch area, and then smooth the surfaces and round the edges, as necessary, using a flexible knife or trowel.

Tape scrap lumber pieces around the patch as a form. Coat the insides with vegetable oil or commercial release agent so the patch won't adhere to the wood. Remove the wood when the patch is firm. Cover with plastic and protect from traffic for at least one week.

How to Patch Step Treads

Make a cut in the stair tread just outside the damaged area using a circular saw with a masonry-cutting blade. Make the cut so it angles toward the back of the step. Make a horizontal cut on the riser below the damaged area, and then chisel out the area in between the two cuts.

Cut a form board the same height as the step riser. Coat one side of the board with vegetable oil or commercial release agent to prevent it from bonding with the repair, and then press it against the riser of the damaged step. Brace it in position with heavy blocks. Make sure the top of the form is flush with the top of the step tread.

Apply latex bonding agent to the repair area with a clean paintbrush, wait until the bonding agent is tacky (no more than 30 min.), and then press a stiff mixture of quick-setting cement into the damaged area with a trowel.

Smooth the concrete with a float, and let it set for a few minutes. Round over the front edge of the nose with an edger. Use a trowel to slice off the sides of the patch, so it is flush with the side of the steps. Cover the repair with plastic and wait a week before allowing traffic on the repaired section.

Miscellaneous Concrete Repairs

There are plenty of concrete problems you may encounter around your house that are not specifically addressed in many repair manuals. These miscellaneous repairs include such tasks as patching contoured objects that have been damaged and repairing masonry veneer around the foundation of your house. You can adapt basic techniques to make just about any type of concrete repair. Remember to dampen concrete surfaces before patching so that the moisture from concrete and other patching compounds is not absorbed into the existing surface. Be sure to follow the manufacturer's directions for the repair products you use.

Tools & Materials ▸

- Putty knife
- Trowel
- Hand maul
- Chisel
- Wire brush
- Aviation snips
- Drill
- Quick-setting cement
- Soft-bristle brush
- Emery paper
- Wire lath
- Masonry anchors
- Concrete acrylic fortifier
- Sand-mix
- Eye protection and work gloves

Concrete slabs that slant toward the house can lead to foundation damage and a wet basement. Even a level slab near the foundation can cause problems. Consider asking a concrete contractor to fix it by mud-jacking, forcing wet concrete underneath the slab to lift the edge near the foundation.

How to Repair Shaped Concrete

Scrape all loose material and debris from the damaged area, and then wipe down with water. Mix quick-setting cement and trowel it into the area. Work quickly—you only have a few minutes before concrete sets up.

Use the trowel or a putty knife to mold the concrete to follow the form of the object being repaired. Smooth the concrete as soon as it sets up. Buff with emery paper to smooth out any ridges after the repair dries.

How to Repair Masonry Veneer

Chip off the crumbled, loose, or deteriorated veneer from the wall using a cold chisel and maul. Chisel away damaged veneer until you have only good, solid surface remaining. Use care to avoid damaging the wall behind the veneer. Clean the repair area with a wire brush.

Clean up any metal lath in the repair area if it is in good condition. If not, cut it out with aviation snips. Add new lath where needed using masonry anchors to hold it to the wall.

Mix fortified sand-mix concrete (or specialty concrete blends for wall repair) and trowel it over the lath until it is even with the surrounding surfaces.

Re-create the surface texture to match the surrounding area. For our project, we used a soft-bristled brush to stipple the surface. To blend in the repair, add pigment to the sand mixture or paint the repair area after it dries.

Brick Repairs

The most common brick wall repair is tuck-pointing, which is the process of replacing failed mortar joints with fresh mortar. Tuck-pointing is a highly useful repair skill for any homeowner to possess. It can be used to repair walls, chimneys, brick veneer, or any other structure where the bricks or blocks are bonded with mortar.

Minor cosmetic repairs can be attempted on any type of wall, from freestanding garden walls to block foundations. Filling minor cracks with caulk or repair compound and patching popouts or chips are good examples of minor repairs. Consult a professional before attempting any major repairs like replacing brick or blocks, or rebuilding a structure—especially if you are dealing with a load-bearing structure.

Tools & Materials ▸

- Raking tool
- Mortar hawk
- Tuck-pointer
- Jointing tool
- Bricklayer's hammer
- Mason's trowel
- Mason's or stone chisel
- Hammer
- Drill with masonry disc and bit
- Pointing trowel
- Stiff-bristle brush
- Mortar (Type S or N)
- Gravel
- Scrap of metal flashing
- Replacement bricks or blocks
- Mortar repair caulk
- Eye protection and work gloves

Make timely repairs to brick and block structures. Tuck-pointing deteriorated mortar joints is a common repair that, like other types of repair, improves the appearance of the structure or surface and helps prevent further damage.

Making Repairs Blend ▸

Add mortar tint to your mortar mix to help repairs blend in. Fresh mortar usually stands out because it is too light. For some repairs, you can use pre-tinted mortar repair caulk.

How to Replace Deteriorated Mortar Joints

Clean out loose or deteriorated mortar to a depth of ¼ to ¾". Use a mortar raking tool (top) first, and then switch to a masonry chisel and a hammer (bottom) if the mortar is stubborn. Clear away all loose debris, and dampen the surface with water before applying fresh mortar.

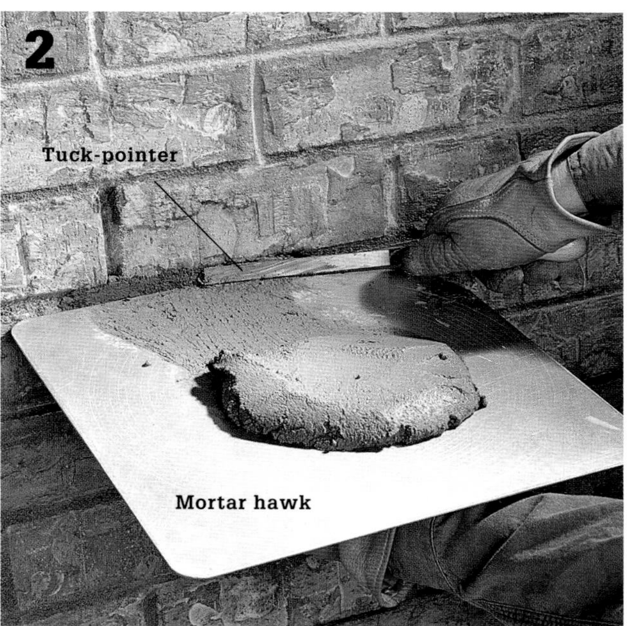

Mix the mortar to a firm workable consistency; add tint if necessary. Load mortar onto a mortar hawk, and then push it into the horizontal joints with a tuck-pointer. Apply mortar in ¼- to ½"-thick layers, and let each layer dry for 30 minutes before applying another. Fill the joints until the mortar is flush with the face of the brick or block.

After the final layer of mortar is applied, smooth the joints with a jointing tool that matches the profile of the old mortar joints. Tool the horizontal joints first. Let the mortar dry until it is crumbly, and then brush off the excess mortar with a stiff-bristle brush.

Apply mortar repair caulk with firm pressure filling the mortar joint from back to front in multiple layers. Use the square applicator tip or a jointing tool to smooth the joint.

(continued)

How to Replace a Damaged Brick

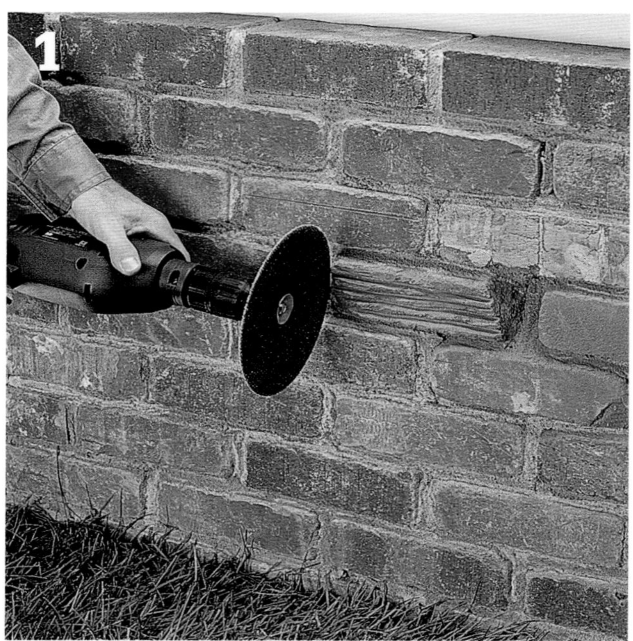

Score the damaged brick so it will break apart more easily for removal. Use a drill with a masonry-cutting disc to score lines along the surface of the brick and in the mortar joints surrounding the brick.

Use a mason's chisel and hammer to break apart the damaged brick along the scored lines. Rap sharply on the chisel with the hammer, being careful not to damage surrounding bricks. *Tip: Save fragments to use as a color reference when you shop for replacement bricks.*

Chisel out any remaining mortar in the cavity, and then brush out debris with a stiff-bristle or wire brush to create a clean surface for the new mortar. Rinse the surface of the repair area with water.

Mix the mortar for the repair and tint if needed to match old mortar. Use a pointing trowel to apply a 1"-thick layer of mortar at the bottom and sides of the cavity.

Dampen the replacement brick slightly, then apply mortar to the ends and top of the brick. Fit the brick into the cavity and rap it with the handle of the trowel until the face is flush with the surrounding bricks. If needed, press additional mortar into the joints with a pointing trowel.

Scrape away excess mortar with a masonry trowel, and then smooth the joints with a jointing tool that matches the profile of the surrounding mortar joints. Let the mortar set until crumbly, and then brush the joints to remove excess mortar.

Tips for Removing & Replacing Several Bricks ▸

For walls with extensive damage, remove bricks from the top down, one row at a time, until the entire damaged area is removed. Replace bricks using the techniques shown above and in the section on building with brick and block. *Caution: Do not dismantle load-bearing brick structures like foundation walls—consult a professional mason for these repairs.*

For walls with internal damaged areas, remove only the damaged section, keeping the upper layers intact if they are in good condition. Do not remove more than four adjacent bricks in one area—if the damaged area is larger, it will require temporary support, which is a job for a professional mason.

Repairing & Replacing Chimney Caps

Chimney caps undergo stress because the temperatures of the cap and chimney flue fluctuate dramatically. Use fire-rated silicone caulk to patch minor cracks. For more extensive repairs, reapply fresh mortar over the cap, or replace the old cap for a permanent solution.

Tools & Materials ▸

- Hammer
- Stone chisel
- Wire brush
- Drill
- Float
- Pointing trowel
- Tape measure
- Caulk gun
- Latex-fortified mortar
- Concrete fortifier
- Plywood (½, ¾")
- ⅜" dowel
- 1½" wood screws
- Vegetable oil or a commercial release agent
- Fire-rated silicone caulk
- Fire-rated rope or mineral wool
- Work gloves

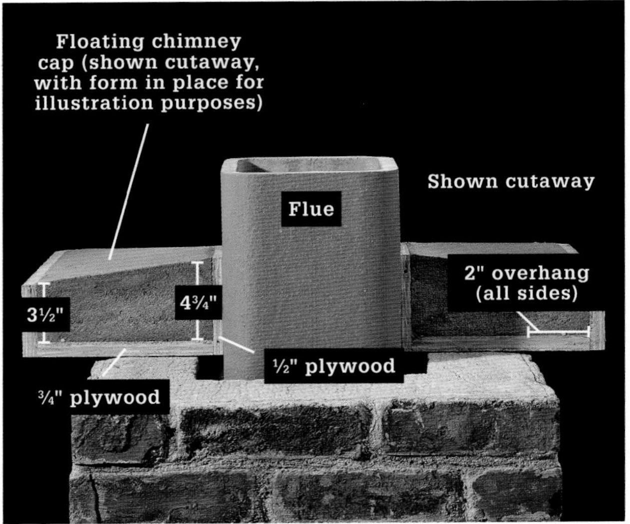

A chimney cap expands and shrinks as temperatures change inside and outside the chimney. This often results in cracking and annual treks up to the roof for repairs. A floating chimney cap (above) is cast in a form using mortar or sand-mix concrete, and then placed on the top of the chimney (opposite page). You can repair a damaged cap by chipping off the deteriorated sections and adding fresh mortar (below).

How to Repair a Chimney Cap

Carefully break apart and remove the deteriorated sections of the chimney cap using a stone chisel and hammer.

Mix a batch of latex-fortified mortar. Trowel an even layer of mortar all the way around the chimney cap, following the slope of the existing cap. Mortar should cover the chimney from the outside edges of the chimney bricks to the flue. Smooth out the mortar with a wood float, trying to recreate the original slope of the chimney cap. Inspect the mortar annually.

How to Cast & Install a Replacement Chimney Cap

Measure the chimney and the chimney flue and build a form from ½ and ¾" plywood (form dimensions on opposite page, top). Attach the form to a plywood base using 1½" wood screws. Glue ⅜" dowels to the base, 1" inside the form. The dowels will cast a drip edge into the cap. Coat the inside of the form with vegetable oil or a commercial release agent.

Prepare a stiff (dry) mixture of mortar to cast the cap—for average-sized chimneys, two 60-lb. bags of dry mix should yield enough mortar. Fill the form with mortar. Rest a wood float across the edges of the form and smooth the mortar. Keep angles sharp at the corners. Let the cap cure for at least a week, and then carefully disassemble the form.

Chip off the old mortar cap completely, and clean the top of the chimney with a wire brush. With a helper, transport the chimney cap onto the roof and set it directly onto the chimney, centered so the overhang is equal on all sides. For the new cap to function properly, do not bond it to the chimney or the flue.

Shift the cap so the gap next to the flue is even on all sides, and then fill in the gap with fire-rated rope or mineral wool. Caulk over the fill material with a very heavy bead of fire-rated silicone caulk. Also caulk the joint at the underside of the cap. Inspect caulk every other year and refresh as needed.

Repairing Stonework

Damage to stonework is typically caused by frost heave, erosion or deterioration of mortar, or by stones that have worked out of place. Dry-stone walls are more susceptible to erosion and popping while mortared walls develop cracks that admit water, which can freeze and cause further damage.

Inspect stone structures once a year for signs of damage and deterioration. Replacing a stone or repointing crumbling mortar now will save you work in the long run.

A leaning stone column or wall probably suffers from erosion or foundation problems and can be dangerous if neglected. If you have the time, you can tear down and rebuild dry-laid structures, but mortared structures with excessive lean need professional help.

Tools & Materials ▸

- Maul
- Chisel
- Camera
- Shovel
- Hand tamper
- Level
- Batter gauge
- Stiff-bristle brush
- Trowels for mixing and pointing
- Mortar bag
- Masonry chisels
- Wood shims
- Carpet-covered 2 × 4
- Chalk
- Compactable gravel
- Replacement stones
- Type M mortar
- Mortar tint
- Eye protection and work gloves

Stones in a wall can become dislodged due to soil settling, erosion, or seasonal freeze-thaw cycles. Make the necessary repairs before the problem migrates to other areas.

Tips for Repairing Popped Stones ▸

Return a popped stone to its original position. If other stones have settled in its place, drive shims between neighboring stones to make room for the popped stone. Be careful not to wedge too far.

Use a 2 × 4 covered with carpet to avoid damaging the stone when hammering it into place. After hammering, make sure a replacement stone hasn't damaged or dislodged the adjoining stones.

How to Rebuild a Dry-stone Wall Section

Study the wall and determine how much of it needs to be rebuilt. Plan to dismantle the wall in a V shape, centered on the damaged section. Number each stone and mark its orientation with chalk so you can rebuild it following the original design. *Tip: Photograph the wall, making sure the markings are visible.*

Capstones are often set in a mortar bed atop the last course of stone. You may need to chip out the mortar with a maul and chisel to remove the capstones. Remove the marked stones, taking care to check the overall stability of the wall as you work.

Rebuild the wall, one course at a time, using replacement stones only when necessary. Start each course at the ends and work toward the center. On thick walls, set the face stones first, and then fill in the center with smaller stones. Check your work with a level and use a batter gauge to maintain the batter of the wall. If your capstones were mortared, re-lay them in fresh mortar. Wash off the chalk with water and a stiff-bristle brush.

Tip for Erosion ▸

If you're rebuilding because of erosion, dig a trench at least 6" deep under the damaged area, and fill it with compactable gravel. Tamp the gravel with a hand tamper. This will improve drainage and prevent water from washing soil out from beneath the wall.

Tips for Repairing Mortared Stone Walls

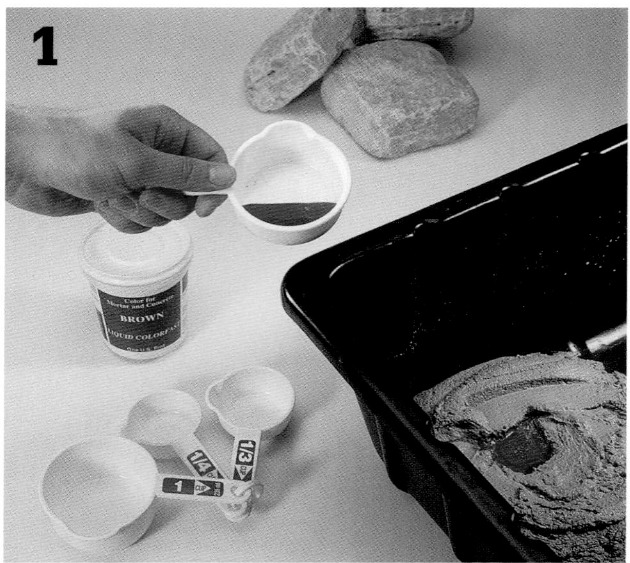

Tint mortar for repair work so it blends with the existing mortar. Mix several samples of mortar, adding a different amount of tint to each and allow them to dry thoroughly. Compare each sample to the old mortar and choose the closest match.

Use a mortar bag to restore weathered and damaged mortar joints over an entire structure. Remove loose mortar (see below) and clean all surfaces with a stiff-bristle brush and water. Dampen the joints before tuck-pointing and cover all of the joints, smoothing and brushing as necessary.

How to Repoint Mortar Joints

Carefully rake out cracked and crumbling mortar, stopping when you reach solid mortar. Remove loose mortar and debris with a stiff-bristle brush. *Tip: Rake the joints with a chisel and maul or make your own raking tool by placing an old screwdriver in a vice and bending the shaft about 45°.*

Mix Type M mortar, and then dampen the repair surfaces with clean water. Working from the top down, pack mortar into the crevices using a pointing trowel. Smooth the mortar when it has set up enough to resist light finger pressure. Remove excess mortar with a stiff-bristle brush.

How to Replace a Mortared Stone Wall

Remove the damaged stone by chiseling out the surrounding mortar using a masonry chisel or a modified screwdriver (opposite page). Drive the chisel toward the damaged stone to avoid harming neighboring stones. Once the stone is out, chisel the surfaces inside the cavity as smooth as possible.

Brush out the cavity to remove loose mortar and debris. Test the surrounding mortar and chisel or scrape out any mortar that isn't firmly bonded.

Dry-fit the replacement stone. The stone should be stable in the cavity and blend with the rest of the wall. You can mark the stone with chalk and cut it to fit (page 179), but excessive cutting will result in a conspicuous repair.

Mist the stone and cavity lightly, and then apply Type M mortar around the inside of the cavity using a trowel. Butter all mating sides of the replacement stone. Insert the stone and wiggle it forcefully to remove any air pockets. Use a pointing trowel to pack the mortar solidly around the stone. Smooth the mortar when it has set up.

Repairing Stucco

Although stucco siding is very durable, it can be damaged and over time it can crumble or crack. The directions given below work well for patching small areas less than two square feet. For more extensive damage, the repair is done in layers, as shown on the opposite page.

Fill thin cracks in stucco walls with a sanded acyrlic stucco caulk. Overfill the crack with caulk and feather until it's flush with the stucco. Allow the caulk to set, and then paint it to match the stucco. Masonry caulk stays semiflexible, preventing further cracking.

Premixed stucco patch works well for small holes, cracks, or surface defects. Repairs to large damaged areas often require the application of multiple layers of base coat stucco and a finish coat stucco. Matching the stucco texture may require some practice.

Tools & Materials ▸

- Caulk gun
- Wire brush
- Putty knife
- Whisk broom
- Hammer
- Masonry chisel
- Drill with masonry bit
- Pointing trowel
- Steel trowel
- Stucco repair products
- Building paper
- Stucco lath
- Roofing nails
- Work gloves, eye protection, and particle mask

Fill thin cracks in stucco walls with stucco repair caulk. Overfill the crack with caulk and feather until it's flush with the stucco. Allow the caulk to set and then paint it to match the stucco. Stucco caulk stays semiflexible, preventing further cracking.

How to Patch Small Areas

Remove loose material from the repair area using a wire brush. Use the brush to clean away rust from any exposed metal lath, and then apply a coat of metal primer to the lath.

Apply premixed stucco patch compound to the repair area, slightly overfilling the hole using a putty knife or trowel.

Smooth the repair with a putty knife or trowel, feathering the edges to blend into the surrounding surface. Use a whisk broom or trowel to duplicate the original texture. Let the patch dry for several days, and then touch it up with masonry paint.

How to Patch Large Areas

Make a starter hole with a drill and masonry bit, and then use a masonry chisel and hammer to chip away stucco in the repair area. Cut self-furring metal lath to size and attach it to the sheathing using roofing nails. Overlap pieces by 2". If the patch extends to the base of the wall, attach a metal stop bead at the bottom. *Note: Wear safety glasses and a particle mask or respirator when cutting stucco.*

Apply a ⅜"-thick layer of base coat stucco directly to the metal lath. Push the stucco into the mesh until it fills the gap between the mesh and the sheathing. Score horizontal grooves into the wet surface using a scratching tool. Let the stucco cure for up to two days, misting it with water every 2 to 4 hours.

Apply a second, smooth layer of stucco after the first coat has become sufficiently firm to hold the second coat without sagging. Build up the stucco to within ¼" of the original surface. Let the patch cure for up to two days, misting every 2 to 4 hours.

Combine finish coat stucco mix with just enough water for the mixture to hold its shape. Dampen the patch area, and then apply the finish coat to match the original surface. Dampen the patch occasionally. Let it cure for 7 days before painting with acrylic or water-based paint and 28 days for oil-based.

Pressure-washing Masonry

A typical residential-grade pressure washer can be as much as 50 times more powerful than a standard garden hose, while using up to 80 percent less water.

A pressure washer comprises an engine to generate power, a pump to force water supplied from a garden hose through a high-pressure hose, and a nozzle to accelerate the water stream leaving the system. This results in a high-pressure water jet ranging from 500 to 4,000 pounds per square inch (PSI).

A pressure washer's cleaning power is noted in gallons per minute (GPM) to rinse away loosened dirt and grime from the area; a pressure washer with a higher GPM cleans faster than a lower-flow unit. For general cleaning around your outdoor home, a pressure washer around 2,500 PSI and 2.5 GPM is more than sufficient.

Pressure washing is quite simple: Firmly grasp the spray wand with both hands, depress the trigger, and move the nozzle across the surface to be cleaned. Although different surfaces require different spray patterns and pressure settings, it is not difficult to determine the appropriate cleaning approach for each project. The nozzle is adjustable—from a low-pressure, wide-fan spray for general cleaning and rinsing, to a narrow, intense stream for stubborn stains. But the easiest way to control the cleaning is to simply adjust the distance between the nozzle and the surface—move the nozzle back to reduce the pressure; move the nozzle closer to intensify it.

To successfully clean any masonry or stone surface using a pressure washer, follow these tips:

- When cleaning a new surface, start in an inconspicuous area, with a wide spray pattern and the nozzle four to five feet from the surface. Move closer to the surface until the desired effect is achieved.
- Keep the nozzle in constant motion, spraying at a steady speed with long, even strokes to ensure consistent results.
- Maintain a consistent distance between the nozzle and the cleaning surface.
- When cleaning heavily soiled or stained surfaces, use cleaning detergents formulated for pressure washers. Always rinse the surface before applying the detergent. On vertical surfaces, apply detergent from bottom to top, and rinse from top to bottom. Always follow the detergent manufacturer's directions.
- After pressure washing, always seal the surface with an appropriate surface sealer (e.g., concrete sealer for cement driveways), following the product manufacturer's instructions.

To clean the masonry and stonework surfaces around the outside of your home, there is nothing that works faster or more effectively than a pressure washer.

Pressure Washer Safety ▸

- Always wear eye protection.
- Wear shoes, but not open-toed shoes.
- Make sure the unit is on a stable surface and the cleaning area has adequate slopes and drainage to prevent puddles.
- Assume a solid stance and firmly grasp the spray gun with both hands to avoid injury if the gun kicks back.
- Always keep the high-pressure hose connected to both the pump and the spray gun while the system is pressurized.
- Never aim the nozzle at people or animals—the high-pressure stream of water can cause serious injury.

Pressure-washing Masonry & Stonework

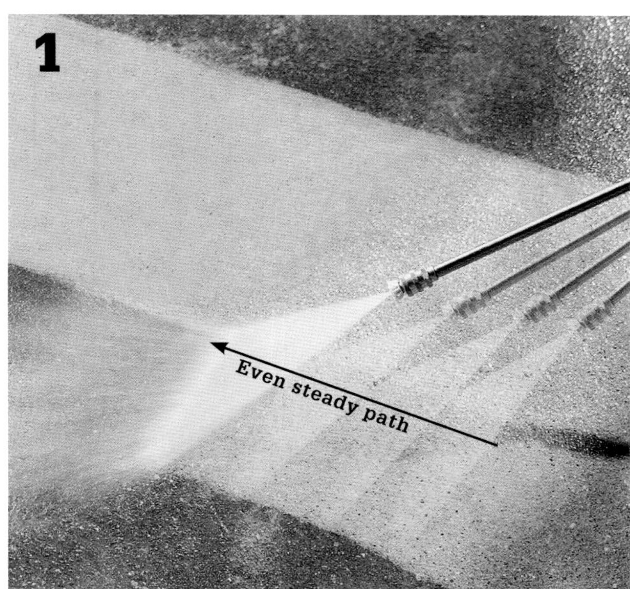

Always keep the nozzle in motion, spraying at a steady speed and using long, even strokes. Take multiple passes over heavily soiled areas. Take care not to dwell on one spot for too long, especially when using narrow, high-pressure spray patterns.

Hold the spray wand so that the nozzle distributes the spray pattern across the surface evenly. Holding the nozzle at too low an angle can cause an uneven spray pattern, resulting in "zebra striping." Also, maintain a consistent distance between the nozzle and the cleaning surface to ensure consistent results and help flush dirt and debris from the area.

Work in identifiable sections, such as the area between the expansion joints in concrete. If there is a slope, work downhill to promote drainage and help flush away dirt and debris. Wet entire surface to prevent streaking.

To prevent streaks on vertical surfaces, always begin pressure washing or applying cleaning detergent at the bottom of the surface, and then work upward. When rinsing, start at the top and work downward—gravity will help the clean water flush away dirt, debris, and detergent residue.

Conversion Table

Converting Measurements

To Convert:	To:	Multiply by:
Inches	Millimeters	25.4
Inches	Centimeters	2.54
Feet	Meters	0.305
Yards	Meters	0.914
Miles	Kilometers	1.609
Square inches	Square centimeters	6.45
Square feet	Square meters	0.093
Square yards	Square meters	0.836
Cubic inches	Cubic centimeters	16.4
Cubic feet	Cubic meters	0.0283
Cubic yards	Cubic meters	0.765
Pints (U.S.)	Liters	0.473 (Imp. 0.568)
Quarts (U.S.)	Liters	0.946 (Imp. 1.136)
Gallons (U.S.)	Liters	3.785 (Imp. 4.546)
Ounces	Grams	28.4
Pounds	Kilograms	0.454
Tons	Metric tons	0.907

To Convert:	To:	Multiply by:
Millimeters	Inches	0.039
Centimeters	Inches	0.394
Meters	Feet	3.28
Meters	Yards	1.09
Kilometers	Miles	0.621
Square centimeters	Square inches	0.155
Square meters	Square feet	10.8
Square meters	Square yards	1.2
Cubic centimeters	Cubic inches	0.061
Cubic meters	Cubic feet	35.3
Cubic meters	Cubic yards	1.31
Liters	Pints (U.S.)	2.114 (Imp. 1.76)
Liters	Quarts (U.S.)	1.057 (Imp. 0.88)
Liters	Gallons (U.S.)	0.264 (Imp. 0.22)
Grams	Ounces	0.035
Kilograms	Pounds	2.2
Metric tons	Tons	1.1

Lumber Dimensions

Nominal - U.S.	Actual - U.S.	Metric
1 × 2	¾ × 1½"	19 × 38 mm
1 × 3	¾ × 2½"	19 × 64 mm
1 × 4	¾ × 3½"	19 × 89 mm
1 × 5	¾ × 4½"	19 × 114 mm
1 × 6	¾ × 5½"	19 × 140 mm
1 × 7	¾ × 6¼"	19 × 159 mm
1 × 8	¾ × 7¼"	19 × 184 mm
1 × 10	¾ × 9¼"	19 × 235 mm
1 × 12	¾ × 11¼"	19 × 286 mm
1¼ × 4	1 × 3½"	25 × 89 mm
1¼ × 6	1 × 5½"	25 × 140 mm
1¼ × 8	1 × 7¼"	25 × 184 mm
1¼ × 10	1 × 9¼"	25 × 235 mm
1¼ × 12	1 × 11¼"	25 × 286 mm
1½ × 4	1¼ × 3½"	32 × 89 mm
1½ × 6	1¼ × 5½"	32 × 140 mm
1½ × 8	1¼ × 7½"	32 × 184 mm
1½ × 10	1¼ × 9½"	32 × 235 mm
1½ × 12	1¼ × 11½"	32 × 286 mm
2 × 4	1½ × 3½"	38 × 89 mm
2 × 6	1½ × 5½"	38 × 140 mm
2 × 8	1½ × 7½"	38 × 184 mm
2 × 10	1½ × 9½"	38 × 235 mm
2 × 12	1½ × 11½"	38 × 286 mm
3 × 6	2½ × 5½"	64 × 140 mm
4 × 4	3½ × 3½"	89 × 89 mm
4 × 6	3½ × 5½"	89 × 140 mm

Liquid Measurement Equivalents

1 Pint	= 16 Fluid ounces	= 2 Cups
1 Quart	= 32 Fluid ounces	= 2 Pints
1 Gallon	= 129 Fluid ounces	= 4 Quarts

Converting Temperatures

Convert degrees Fahrenheit (F) to degrees Celsius (C) by following this simple formula: Subtract 32 from the Fahrenheit temperature reading. Then mulitply that number by 5⁄9. For example, 77°F - 32 = 45. 45 × 5⁄9 = 25°C.

To convert degrees Celsius to degrees Fahrenheit, multiply the Celsius temperature reading by 9⁄5, then add 32. For example, 25°C × 9⁄5 = 45. 45 + 32 = 77°F.

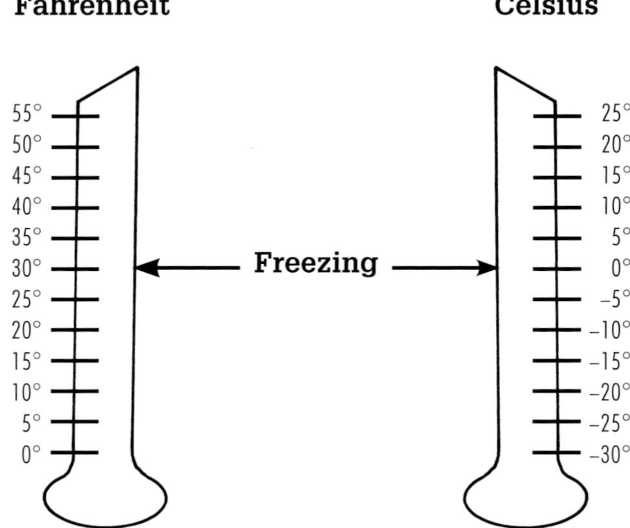

Resources

Anchor Wall Systems
877-295-5415
www.anchorwall.com

Becker Architectural Concrete
651-554-0346
www.beckerconcrete.com

Belgard Pavers
800-899-8455
www.belgardpavers.com

Black & Decker
Portable power tools and more
www.blackanddecker.com

Cultured Stone Corporation
800-255-1727
www.culturedstone.com

Estudio Arqué
+ 34-956-695-896 or + 34-690-656-675
email: info@estudioarque.com
www.estudioarque.com

Kemiko Concrete Floor Stain
903-587-3708
www.kemiko.com

NovaBrik
866-678-BRIK (2745)
www.novabrik.com

Quikrete
Cement and concrete products
800-282-5828
www.quikrete.com

Red Wing Shoes Co.
Work shoes and boots shown throughout book
800-733-9464
www.redwingshoes.com

Seattle Glass Block
425-483-9977
www.seattleglassblock.com

Stanley Tools
800-262-2161
www.stanleytools.com

U-Line
414-354-0300
www.u-line.com

Photography Credits

Anchor Block
p. 9 (lower right)

Belgard
p. 9 (top right), 10 (lower left), 11 (lower), 12 (lower), 15 (top two)

Cultured Stone Corp.
p. 256

Derek Fell's Horticultural Library
p. 9 (lower left)

Eldorado
p. 11 (top)

Tony Giammarino
p. 227 (lower)

iStock / www.istockphoto.com
p. 8 (both), 170, 211 (both), 234

Clive Nichols
p. 80 Hampton Court 2002 / Mercedes-benz. Designer: Jane Monney: Bright blue wall and raised bed planted with herbaceous plants. Sculpture by John Brown

Novabrik
p. 262

Jerry Pavia
p. 6, 9 (top left), 15 (lower), 169, 188, 206, 207 (lower two), 226, 232

Photolibrary / www.photolibrary.com
10 (lower middle) Andrew Lord, 10 (lower right), 12 (top), 13,14

Room & Board
p. 5 (top right), 244

Beth Singer
p. 250

Index

R

S

CREATIVE PUBLISHING international

NOTE TO READERS

The DVD disk included with this book is offered as a free premium to buyers of this book.

The live video demonstrations are designed to be viewed on electronic devices suitable for viewing standard DVD video discs, including most television DVD players, as well as a Mac or PC computer equipped with a DVD-compatible disc drive and standard multi-media software.

In addition, your DVD-compatible computer will allow you to read the electronic version of the book. The electronic version is provided in a standard PDF form, which is readable by any software compatible with that format, including Adobe Reader.

To access the electronic pages, open the directory of your computer's DVD drive, and click on the icon with the image of this book cover.

The electronic book carries the same copyright restrictions as the print version. You are welcome to use it in any way that is useful for you, including printing the pages for your own use. You can also loan the disc to friends or family members, much the way you would loan a printed book.

However, we do request that you respect copyright law and the integrity of this book by not attempting to make electronic copies of this disc, or by distributing the files electronically via the internet.